Eric Bird

Coastal Cliffs: Morphology and Management

AF793692

Springer

Eric Bird
Department of Geography
University of Melbourne
Melbourne, VIC
Australia

and

Geostudies Pty Ltd.
Black Rock, VIC
Australia

Every effort has been made to contact the copyright holders of the figures and tables which have been reproduced from other sources. Anyone who has not been properly credited is requested to contact the publishers, so that due acknowledgment may be made in subsequent editions.

ISSN 2191-5369 ISSN 2191-5377 (electronic)
SpringerBriefs in Earth Sciences
ISBN 978-3-319-29083-6 ISBN 978-3-319-29084-3 (eBook)
DOI 10.1007/978-3-319-29084-3

Library of Congress Control Number: 2016930286

© The Author(s) 2016
This work is subject to copyright. All rights are reserved by the Publisher, whether the whole or part of the material is concerned, specifically the rights of translation, reprinting, reuse of illustrations, recitation, broadcasting, reproduction on microfilms or in any other physical way, and transmission or information storage and retrieval, electronic adaptation, computer software, or by similar or dissimilar methodology now known or hereafter developed.
The use of general descriptive names, registered names, trademarks, service marks, etc. in this publication does not imply, even in the absence of a specific statement, that such names are exempt from the relevant protective laws and regulations and therefore free for general use.
The publisher, the authors and the editors are safe to assume that the advice and information in this book are believed to be true and accurate at the date of publication. Neither the publisher nor the authors or the editors give a warranty, express or implied, with respect to the material contained herein or for any errors or omissions that may have been made.

Printed on acid-free paper

This Springer imprint is published by SpringerNature
The registered company is Springer International Publishing AG Switzerland

SpringerBriefs in Earth Sciences

More information about this series at http://www.springer.com/series/8897

Preface

Coastal cliffs have formed as the result of erosion of the borders of uplands. They present a range of problems, many of which are the result of their instability: they have retreated and are still receding, some rapidly, others very slowly. As they recede, areas of land are consumed, most of which are valued, either because of their utilisation for agriculture, industry, human settlements or communication, or because they are resources of scenic, recreational or scientific significance.

Coastal cliffs are landforms produced and modified by geomorphological processes. Their abruptness provides secure habitats for plants and animals, particularly birds, but they are also hazardous, a place where accidents happen; where people can come to harm. They provide sections exposing geological formations and are consequently of educational value and features that stimulate research in geology, geomorphology and ecology: features and processes. They offer opportunities for recreation, such as climbing and abseiling, or launching places for kites and hang-gliders. They attract artists and photographers, walkers, bird watchers, as well as people who simply come to enjoy the view. Their scenic attraction and ecological importance have been recognised in the designation of coastal reserves of various kinds, including national parks, nature reserves, sites of scientific interest, bird sanctuaries and recreation areas.

This book provides an introduction to the geology and geomorphology of cliffs, their evolution and the changes taking place on them. There is consideration of the various coastal processes that have shaped them and that result in changes. There is explanation of how they change, and the differing rates at which they recede. On some coasts, cliff recession has been halted, but this is not always necessary or desirable. Where they have been stabilised in on way or another there have sometimes been unexpected consequences, such as the loss of nearby beaches. The decision on whether to stabilise cliffs or allow them to continue to recede is an important part of coastal cliff management.

Such management should be based on an understanding of geomorphology: how a cliff or bluff has formed, how it is changing naturally and as the result of human activities and how management procedures are likely to modify it. This

requires knowledge of the processes at work, their effects and changes likely to occur whether they are modified. The present book provides background to this knowledge.

The text is accompanied by numerous photographs, intended to illustrate and clarify concepts of cliffed coasts and their management. The author is of the opinion that these illustrations provide important supplements to the text, showing the variety of situations that have arisen on coasts around the world. Lists of References provide a guide to more detailed information, and a Glossary is provided to explain the meanings of technical terms used in the text.

Eric Bird is a Principal Fellow in the Department of Geography, University of Melbourne, and Director of Geostudies. He is the author of several books, including *Beach Management* (1996), *Coastal Geomorphology: An Introduction* (1999, 2008), *Submerging Coasts* (1993), an *Encyclopedia of the World's Coastal Landforms* (2010) and (with Nick Lewis) *Beach Renourishment* (2015). He lives and writes in Melbourne.

Contents

Chapter 1
Introduction

Abstract This chapter defines cliffs and bluffs and outlines where they occur on the world's coastlines. They have been cut in various geological formations, and have taken shape as a sequel to a world-wide sea level rise.

This book is intended for people interested the evolution and dynamics of coastal cliffs, efforts to preserve their scenery and scientific interest, where it is necessary to stabilise them, and how to ensure the safety of people who visit them. Numerous colour illustrations are provided to show the great diversity of coastal cliffs around the world and the variety of problems posed.

This chapter describes coastal cliffs, their global distribution and geological relationships. Chapter 2 discusses their geomorphological evolution in relation to coastal processes, and the shaping of coastal profiles. Chapter 3 describes the various processes at work on coastal cliffs, and the effects of human activities. Chapter 4 examines varying rates of cliff recession and their causes. Chapter 5 deals with hazards associated with cliffs and what can be done about them. Chapter 6 explores scenic and scientific values on cliffed coasts, and demands for their preservation or various kinds of development. Chapter 7 discusses means of stabilising coastal cliffs where this is necessary, and Chap. 8 summarises the need for knowledge of the processes at work on cliffs, their effects, and the changes likely to occur if they are modified. A Glossary of terms that may not be familiar to the reader is provided at the end of the book.

Coastal cliffs are steep slopes exposing rock formations, and are found wherever high ground has been intersected by a retreating coastline (Guilcher 1958; Sunamura 1992). They generally have gradients of more than 40°, and are often vertical and sometimes overhanging. They occupy many parts of the world's coasts, and their geomorphology is strongly influenced by geology, particularly the structure and lithology of the outcropping rock formations. Most cliffs have receded as the result of erosion, but some are now stable. Where they are actively receding, consuming useful agricultural or forested land or threatening coastal structures such as roads, buildings (Fig. 1.1) and seaside towns, it may be necessary to halt their retreat, or at least to modify it, by coastal cliff management.

© The Author(s) 2016

E. Bird, *Coastal Cliffs: Morphology and Management*,
SpringerBriefs in Earth Sciences, DOI 10.1007/978-3-319-29084-3_1

Fig. 1.1 Cliff recession becomes a problem when it results in the undermining and destruction of buildings, as in this example on the Cumbrian coast in NW England. © Geostudies

It should be borne in mind that a retreating cliff is a source of sediment for nearby beaches, and that the stabilisation of a cliff, notably by the construction of a basal sea wall, cuts off this sediment supply. One of the several possible causes of beach erosion is the stabilisation of cliffs that previously yielded a sediment supply to the shore (Bird and Lewis 2014, pp. 9–11).

Coastal cliff management requires an understanding of how a cliff formed and of changes taking place on it as the result of coastal processes. It usually aims to halt cliff recession, or at least reduce it to an acceptable rate.

Some coastal cliffs are changing very slowly if at all, and if they acquire a cover of soil and vegetation they are termed bluffs. There is a difficulty with this term because it has sometimes been used to describe coastal cliffs: in North America, particularly, the term bluff can describe a slope that is devoid of vegetation and subject to erosion. Steep vegetated coastal slopes are generally stable, but may have occasional landslides, which can result in coastline recession; a situation that exists on the steep forested coasts of Washington and Oregon as well as on many tropical coasts. Bluffs that are intermittently unstable may also require management to reduce or prevent the loss of valuable land resources, including coastal roads and railways.

Cliffs 100–500 m high are termed high cliffs, while those exceeding 500 m are megacliffs (Guilcher 1966): an example is the megacliff on Vidoy in the Faerø Islands, 725 m high. Microcliffs are generally less than a metre high, formed where marine processes have attacked the margins of low-lying coastlands.

Cliffs and steep coasts are extensive on the Pacific seaboards of North and South America and in Argentina and southern Brazil. They occur in Venezuela and Columbia, and around many Caribbean islands, and are found on the east coast of North America north from New Jersey, including Labrador. They border much of Greenland and Iceland, Scandinavia and parts of the Baltic coasts of Poland, Germany and Denmark. There are cliffs and steep coasts in the British

Isles (Clayton et al. 2003), and in France, Spain and Portugal, and around much of the Mediterranean and Black Seas. They occur sporadically around Africa and Arabia and along the coasts of Iran, Pakistan and India, and locally interrupt the generally low-lying coasts of south-east Asia, China, and the Koreas. Much of the Pacific coast of Russia is steep and cliffed, and the Russian Arctic coast has some high parts, as does the northern coast of Canada. Indonesia and Papua New Guinea have sectors of steep and cliffed coast, and they occur widely in Australia and New Zealand and around oceanic islands. Much of Antarctica is ice-fringed, but there are a few steep and cliffed coasts as well as some rocky shores, exposed to marine processes at least during the summer, after shore ice has thawed.

Many coastal cliffs border headlands separated by low-lying embayments, but there are also long cliffy sectors, as on the Nullarbor coast of southern Australia, and lengthy stretches of steep coast, as in Oregon and Washington. Around the world cliffs have been cut into rock formations ranging in age from Pre-Cambrian (more than 560 million years old) in northern Canada to Pleistocene glacial drift (deposited only tens of thousands of years ago) on the Polish coast in the southern Baltic Sea. Most coastal cliffs attained their present form during the Holocene, particularly during the past 6000 years, when the sea has stood at or close to its present level as a sequel to the world-wide Late Quaternary marine transgression.

As a result of this marine transgression (sea level rise) some steep slopes became partially submerged plunging cliffs (Fig. 1.2), but generally there has been erosion, most cliffs having been shaped by marine processes operating with the sea at or close to its present level. A few steep coasts retain features such as 'raised beaches' and emerged rocky shorelines, formed when the sea stood at a higher level relative to the land. Other coastal slopes include features shaped by subaerial processes during episodes when the sea was lower, and not yet removed by marine processes. For example, some coasts retain features that were shaped by glacial and periglacial processes during cold periods in the Pleistocene (§2.4): on some coasts in high latitudes these processes are still active, at least in winter. In temperate and tropical environments there are some coasts that still have features shaped by subaerial processes under preceding warmer and perhaps wetter

Fig. 1.2 Steep coastal slope and plunging cliff on Slieve League, Donegal, Ireland. This megacliff (598 m high) consists of steep (30°–60°) scree-clad slopes that descend to vertical plunging cliffs 100–200 m high. © Geostudies

climatic conditions, and there are coasts in low latitudes where these processes are still active. Cliffs and steep coasts that retain such inherited topographic features have clearly not receded far, for they would be destroyed by subsequent cliff recession.

References

Bird ECF, Lewis N (2014) Beach renourishment. Springer, Dordrecht

Clayton KM, Bird ECF, Hansom JD (2003) The geomorphology of the coastal cliffs of Great Britain. In: May VJ, Hansom JD (ed) Coastal geomorphology of Great Britain. Geological conservation review series 28, Joint Nature Conservation Committee, Peterborough, pp 33–51

Guilcher A (1958) Coastal and submarine morphology. Methuen, London

Guilcher A (1966) Les grandes falaises et megafalaises des côtes sud-ouest et ouest d'Irlande. Annals de Géographie 75:26–38

Sunamura T (1992) Geomorphology of rocky coasts. Wiley, Chichester

Chapter 2
Evolution of Cliffs

Abstract Cliffs are related to geological structure and rock types, and have evolved in response to tectonic movements of the land and changes in sea level. Their profiles are influenced by rock resistance, hinterland topography, exposure to wave action and the effects of coastal processes. Steep vegetated coastal slopes, termed bluffs, are generally former cliffs degraded by weathering; most are stable, but some are subject to occasional landslides. Slope-over-wall coasts occur where a weak formation (such as clay) rests upon more resistant rock, or where a coastal slope has been undercut by marine erosion; in high latitudes the slope may be inherited from prior weathering during periods of cold climate.

2.1 Origin of Cliffs

Cliffs border coastal uplands, produced by uplift of the land and lowering of sea level through geological time; for example there are extensive cliffs cut in Chalk, a rock formation of marine origin that originated in the Cretaceous Sea, 65 to about 100 million years ago, but is now above sea level. Most cliffs are on coasts that have been stable during the past few thousand years, but some have been modified by tectonic uplift or sea level fluctuations within this period. The simplest cliffs are found where marine erosion has attacked the margins of a stable land mass, removing a wedge of material to form a steep slope that may be vertical or even overhanging, fronted by a seaward-sloping shore platform that extends from the cliff base out to below low tide level (Fig. 2.1).

A few coastal cliffs originated as the result of uplift of the land along a fault or monocline and accompanying subsidence on the down-thrown side (Fig. 2.2). Slopes formed by dislocation along a fault line (so that they coincide with the fault plane) are known as fault scarps, and cliffs that originated in this way may be termed fault cliffs. They are found on parts of the Mediterranean coastline, as on the steep and cliffed coasts of Malta (Fig. 2.3), which have formed at least partly

© The Author(s) 2016
E. Bird, *Coastal Cliffs: Morphology and Management*,
SpringerBriefs in Earth Sciences, DOI 10.1007/978-3-319-29084-3_2

Fig. 2.1 A vertical Chalk cliff and seaward-sloping intertidal shore platform at Birling gap on the Sussex coast. © Geostudies

Fig. 2.2 Cliff of Miocene sandstone in Beaumaris Bay, Melbourne Australia, following the axis of the Beaumaris monocline. © Geostudies

Fig. 2.3 The steep coast at Dingli in Malta, probably initiated by faulting, now consists of a rubble-strewn slope descending to a vertical cliff cut in Miocene limestone. A structural bench is seen at the base of the cliff. © Geostudies

as the result of land uplift along fault lines, which produced steep slopes that have been only slightly modified by marine erosion and weathering (defined as the disintegration or decomposition of a rock surface exposed to the atmosphere as the result of physical, chemical and biological processes). Cliffs formed as the result of uplift along a fault usually took shape as the result of repeated tectonic uplifts (earthquakes) on a small scale, each producing a metre or so of exposed fault plane. Recurrent earthquakes can form a high cliff, with several stages of exposure of the fault plane (their age diminishing downward), but weathering and erosion have modified the earlier stages of such a composite cliff, so that only the latest stage shows a freshly-exposed fault plane.

Some cliffs that follow fault lines are the outcome of differential erosion, where faulting had previously placed soft rock formations alongside more resistant rocks and removal of the softer rock has exposed the fault plane as a cliff. In fact very few coastal cliffs follow the lines of active or inactive faults, although some have been cut across transverse fault lines (Fig. 2.4).

Examples of a cliff produced by land uplift can also be seen where a fringing coral reef has been raised out of the sea (Fig. 2.5). Many of these have been modified by erosion and solution, forming 'notch-and-visor' profiles (Fig. 2.6). In the Loyalty Islands, New Caledonia, coral atolls have been raised and tilted by earth movements near the margins of a tectonic plate, forming cliffs with several notched terraces, each cut by the sea during an interval of stability between phases of intermittent uplift (Fig. 2.7).

A few cliffs have originated as the result of volcanic eruptions. Examples include the steep cliffs facing into the crater of Krakatau in Sunda Strait, Indonesia, formed by the explosive eruption of that volcano in 1883 (Fig. 2.8). Similar cliffs are seen on the Greek volcanic island of Santorini in the Mediterranean Sea, and on Deception Island in the South Shetland Islands, Antarctica.

Fig. 2.4 The sloping cliff (*left*) in Rinsey Cove, Cornwall, is a fault-line cliff exposed along the exhumed contact between the Mylor Slates and the overthrust Tregonning Granite, both of Devonian age. © Geostudies

Fig. 2.5 Notched coastal slope on emerged coral limestone at Baron, on the south coast of Java, Indonesia. © Geostudies

Fig. 2.6 Notch-and-visor on a cliff of emerged coral, Isle of Pines, New Caledonia. © Geostudies

Fig. 2.7 Multiple terraces formed during stages in intermittent uplift on the tilted coral island of Uvéa, Loyalty Islands, New Caledonia. © Geostudies

Fig. 2.8 The steep cliff on the island of Rakata is the wall of a volcanic crater formed by the 1883 explosion of Krakatau in Sunda Strait Indonesia. © Geostudies

2.2 Cliffs Formed by Marine Erosion

Most cliffs have been formed mainly by marine erosion of the margins of a pre-existing upland during the Holocene: generally the few thousand years since the Late Quaternary marine transgression brought the sea to its present level.

Waves breaking on the shore are powerful agents of abrasion (the wearing of a rock surface by friction, also known as corrasion) as the result of the hydraulic pressure of wave impact, which forces air into crevices and so generates pressure alternations that widen them. Abrasion is intensified where the waves mobilise rock fragments (sand, gravel or boulders), which are hurled at the cliff base during storms. This causes wearing, grinding and scraping, and can result in quarrying of rock fragments or excavation of sediment. Such erosion leads to the cutting of a sharp-angled junction or an undercut abrasion notch at the cliff-base (Fig. 2.9). Formation of a cliff-base notch is followed by the collapse of rock from the cliff face and the accumulation of rocky debris as talus at the base of the cliff. This is gradually consumed or removed by wave action until waves can again attack the base of the cliff.

Fig. 2.9 Cliff-base notch at Birling Gap, Sussex, formed largely by abrasion, which has cut across flint horizons in the Chalk. © Geostudies

2.2.1 Cliff Profiles and Lithology

The effects of marine erosion depend on the nature and resistance of the rock formations encountered. Their lithology varies from very hard quartzites, igneous formations such as granites and indurated metamorphic rocks such as greenstones, through moderately hard slates to sandstones and soft limestones such as chalk and weak mudrocks such as shales and clays. Categories of rock hardness suggested by Clayton and Shamoon (1998) were:

- Very hard quartzites and sandstones, massive granite and indurated metamorphic rocks.
- Moderately hard slates, shales, grits and basalts.
- Weak limestones (including Chalk) and sandstones.
- Very weak mudrocks (defined as rocks containing at least 90 % silt and clay, and formerly known as siltstones, shales and clays) and unconsolidated sands.

The resistance of an exposed rock formation depends not only on its hardness, but also on its structure, notably planes of division such as joints and bedding planes and lines or zones of relative weakness that are exploited by erosion and weathering processes.

Cliff profiles on resistant rock outcrops are generally bold and those on weak formations gentler: many of the bold headlands on the coast of Cornwall are on outcrops of dolerite (greenstone), in contrast with gentler coastal slopes on slates or phyllites (Bird 1998). In general, cliffs on rock outcrops on older geological formations (such as Pre-Cambrian and Palaeozoic outcrops on the Atlantic coasts of Britain) recede slowly because these rocks are more resistant to weathering than younger formations, but there are many exceptions to this: in northern Scandinavia and northern Canada there are cliffs cut into horizontally-bedded, relatively soft unconsolidated strata of Cambrian age, while cliff outcrops on Cretaceous and Tertiary rocks bordering the Mediterranean Sea have locally been hardened by intense tectonic compression during the Alpine earth movements. The Chalk cliffs at The Needles, at the western end of the Isle of Wight, persist in rock that has been made more resistant by intense folding.

Marine erosion has had little effect on massive resistant formations such as granite, which sometimes forms a steep coastal slope descending into the sea instead of cliffs and shore platforms (Fig. 2.10). This is the outcome of a long period of slow subaerial weathering. Typically, however, granite cliffs show dissection along joints (Fig. 2.11), and the cliffs at Lands End in Cornwall have a castellated appearance related to the excavation of clefts along vertical, inclined and horizontal joints (Fig. 2.12). Cliff profiles on hard sandstone (quartzite) may be similar. The Old Red Sandstone is a resistant formation that forms high cliffs up to 60 m high on the north coast of Scotland between Dunscansby and Skirza Head and on the coast of the Orkney Islands; it is stratified and jointed, and cliff profiles are influenced by erosion along these planes of division (Fig. 2.13). Consolidated volcanic rocks such as basalt also show dissection along planes

Fig. 2.10 Granite coast south of Cape Tourville, Tasmania where the granite slopes down into the sea. © Geostudies

Fig. 2.11 The influence of jointing on the profile of a granite cliff in the fiord of Qeqertarsuatsiat, SW Greenland. © Geostudies

Fig. 2.12 Pendower Cove, cliffs and caves in castellated granite near Land's End, Cornwall. © Geostudies

Fig. 2.13 The cliffs at Yesnaby on Orkney, dissected along joint planes in the Old Red Sandstone. © Geostudies

Fig. 2.14 Columnar basalt cliff at Fingal Head, New South Wales. © Geostudies

between successive lava flows or along joints, which may separate columnar structures as in the Giant's Causeway in Northern Ireland, Fingal Head in New South Wales (Fig. 2.14) and the vertical cliffs and stepped rocky shores on columnar dolerites at Pillar Point in south-east Tasmania. Such volcanic columns are usually vertical, but may have been modified by tectonic movements (Fig. 2.15).

On rocks of moderate resistance, such as Chalk, wave action undercuts the cliff, leading to the collapse of rock from the cliff face, which remains vertical or steep. There is often a basal abrasion ramp that declines to a gently-sloping or horizontal shore platform, often influenced by the outcrops of rock strata (Fig. 2.16). Cliffs in moderately resistant sandstone cliffs show variations related to sedimentary texture and structure, notably folding and faulting, and joints and bedding-planes (Figs. 2.17).

On soft rock formations vertical cliffs have been cut into coherent, homogeneous formations such as the Pleistocene brickearth (loess) of Pegwell Bay in Kent and the pindan deposits of the Dampier Peninsula in the Broome district of NW

Fig. 2.15 Cliff in basaltic columns that have been tilted and deformed by earth movements. © Geostudies

Fig. 2.16 Chalk cliff and shore platform at Vaucottes, northern France, cut into horizontally bedded strata. © Geostudies

Australia (Fig. 2.18). Pindan is a soft, coherent red silty sand of aeolian origin, winnowed from a desert hinterland during dry phases in the Pleistocene. It sustains a vertical outcrop in coastal cliffs, which retreat as the result of recurrent slumping, remaining vertical as the slumped talus is dispersed by wave action. Pindan is much like the European loess, which is a silty aeolian deposit that formed during glacial periods (blown from a cold, dry morainic landscape), and is seen in vertical cliffs on the NW coast of the Black Sea.

On other soft formations, such as clay or sand, marine erosion is accompanied by subaerial weathering and erosion that results in slumping, and the formation of a wide irregular coastal slope (Fig. 2.19). Gulleying and slumping are seen on cliffs cut in glacial drift, as on the coasts of Block Island, Martha's Vineyard and Nantucket Island in Massachusetts, where the cliffs are receding at up to a metre per year, and on the Holderness cliffs in Yorkshire (Fig. 2.20), where cliffs cut in glacial

Fig. 2.17 Vertical cliff in stratified Jurassic sandstone with almost horizontal bedding-planes, West Bay, Dorset, England. © Geostudies

Fig. 2.18 Vertical cliffs in pindan deposits on the coast near Broome, Western Australia, showing disintegration along joints. © Geostudies

drift have been retreating for several centuries: many villages and farms have been lost. Similar slumping slopes, partly grassed, occur on the London Clay outcrop at Warden Point, on the Isle of Sheppey in Kent (Swale Borough Council 2011).

Recurrent slumping has also occurred, particularly after wet weather or the thawing of a snow cover, on the soft Tertiary sands and clays of the Bournemouth coast and the northern shores of the Isle of Wight, on Jurassic clays and shales on the Dorset and Yorkshire coasts, and on glacial drift deposits, as in eastern England, the Danish archipelago, New England or the islands of Puget Sound. Profiles of receding cliffs cut in glacial drift on the North Norfolk coast between West Runton and Sheringham, are bold on clay outcrops and sloping and slumping where the glacial drift is sandy.

On recently-formed volcanic islands (such as Surtsey, which started to form in 1963 off the south coast of Iceland) the softer sediment (volcanic ash, also known

Fig. 2.19 Successive landslides have led to rapid recession of a cliff at Blackgang Chine, Isle of Wight, behind a slumping undercliff on soft Lower Cretaceous clays and sands. © Geostudies

Fig. 2.20 Retreating cliffs cut in Pleistocene boulder clay on the Holderness coast in Yorkshire. © Geostudies

as tuff) is soon removed by erosion, but where hard lava outcrops at or above sea level, cliffs predominate.

2.2.2 Cliff Profiles and Geological Structure

Cliff morphology is also influenced by geological structure: massive rock outcrops form steep coasts, but cliff profiles on stratified or fractured rock formations are related to the dip, joint patterns, bedding planes, cleavage planes, faults and folds and the disposition of harder and softer components. Solid and massive formations are generally eroded more slowly than formations that disintegrate readily along joints, bedding planes, cleavage planes and fractured zones that facilitate

Fig. 2.21 Chalk forms vertical cliffs cut into strata that dip gently seaward at the Seven Sisters, Sussex. © Geostudies

cliff dissection. Most rock formations have planes of division that are weakened by weathering processes and penetrated by marine erosion, influencing the outline of a cliff and forming crevices, clefts, inlets and caves along the coast.

Where the dip is seaward the cliff profile remains steep because dislodged rock slides away. On the coasts of Kent and Sussex Chalk forms vertical cliffs cut into strata that dip gently seaward (Fig. 2.21), and landslides have occurred where the dipping Chalk rests upon Gault clay, as at Folkestone Warren (§3.5). On some coasts evenly-sloping cliffs may follow an exposed seaward-dipping bedding plane in stratified rock. Vertical cliffs may be formed where joint planes are exposed, as on the Lower Cretaceous sandstone cliffs at Hastings in Sussex (Fig. 2.22). Figure 2.23 shows structural ledges as steps on horizontal sandstones separated by weak clays.

Fig. 2.22 Angular pattern in cliffs in sandrock near Hastings in Sussex, related to exposure of vertical joint planes. © Geostudies

Fig. 2.23 Structural benches of Devonian sandstone separated by shale on Bay cliff at Wonboyn, south coast New South Wales. © Geostudies

Where the Chalk outcrop dips steeply and is often strongly folded, as in the Isle of Wight and on the Dorset coast, there are more irregular features related to minor variations in lithology, such as ledges and reefs on the more resistant layers (e.g. the Melbourn Rock) and gentler slopes on softer marly horizons within the Chalk.

Where the dip is landward, as at Ballard Down in Dorset (Fig. 2.24) the cliffs undercut the steep slopes of the Chalk escarpment, and may be described as escarpment cliffs (Bird 1995).

On the north coast of Devon the cliffs near Hartland Point are generally vertical, cut back by stormy seas across Carboniferous sandstones and shales compressed into tight zigzag folds along vertical axes running at right angles to the coastline. In detail the harder sandstones protrude as ribs or small salients and the weaker shales have been cut out as coves, while on a larger scale there are ridges on anticlinal zones (upfolds) that have proved slightly more resistant, and inlets along the transverse synclines (downfolds).

Cliff profiles are thus related to variations in rock resistance, picked out by erosion as the cliff is cut back. The more resistant parts of coastal rock formations protrude as ledges, or persist as caprock on rocky stacks and islands offshore,

Fig. 2.24 Escarpment cliff at Ballard Down, Dorset. © Geostudies

Fig. 2.25 Cliffs in Carboniferous Limestone, Flimston Bay, Pembrokeshire. © Geostudies

whereas the weaker elements are cut back as cliffs and caves. Weathering and erosion penetrate zones of weakness such as faults, joints or outcrops of less resistant rock, cutting clefts and crevices that may develop into caves and blowholes or deep, narrow inlets, and isolating stacks. Cliffed coasts showing an array of such features are found on Carboniferous Limestone in south Pembrokeshire (Fig. 2.25), while on the Normandy coast of France the Porte d'Aval is a slender natural arch on the cliffs of hard (silicified) Chalk at Etretat (Fig. 2.26).

Sloping cliffs, sometimes vegetated, occur where a resistant formation (e.g. hard limestone) overlies a weaker horizon (e.g. soft sandstone) outcropping down to sea level, as at Quobba on the west coast of Australia. On the south-west coast of the Isle of Wight sandstone horizons in the clay-dominated Wealden Beds produce relatively steep cliffs where they outcrop at the cliff crest, as at Barnes High, and cliff-face ledges where they outcrop in the cliff profile, as at Sudmoor Point (Bird 1997). Steep, receding cliffs in Lias Clay are seen on the Dorset coast at Seatown (Fig. 2.27).

Contrasts in lithology have influenced the profiles of cliffs cut in Jurassic formations along the Lyme Bay coast in Dorset (Bird 1995). On Golden Cap

Fig. 2.26 Cliffs and stacks in thinly-bedded silicified Chalk at Etretat, on the Normandy coast in France. © Geostudies

Fig. 2.27 Sloping cliff in Lias clay at Seatown, Dorset. © Geostudies

Fig. 2.28 Golden Cap, Dorset, the highest cliff on the south coast of England rising 191 m above sea level, with a capping of Upper Greensand over Lower Jurassic (Lias) clays and limestone layers. © Geostudies

(Fig. 2.28) the harder sandstone outcrops form steep slopes and ledges and the softer clays gentler (partly vegetated) slopes and areas of subsidence. The high cliffs on the south coast of Malta are vertical on the Lower Coralline Limestone but sloping on the overlying Globigerina Limestone (Fig. 2.3).

Some cliffs show vertical grooves and buttresses cut out along joint planes. Examples are seen on soft Triassic sandstone at High Peak on the south coast of Devon (Fig. 2.29) and on the Chalk cliffs near the mouth of the Cuckmere River in Sussex (Fig. 2.30). Vertical grooves and buttresses have been excavated in the cliffs of Alum Bay, on the Isle of Wight, cut in alternations of Eocene sandstone and clay strata that stand vertically on the northern side of a monoclinal fold.

Fig. 2.29 Buttresses and grooves in sandstone on the cliff at High Peak, near Sidmouth on the south coast of Devon. © Geostudies

Fig. 2.30 Buttresses on the Chalk cliffs near Cuckmere Haven, Sussex. © Geostudies

2.2.3 Cliff Profiles and Exposure to Wave Action

Cliffs on coasts exposed to strong wave action from the open sea are typically bold, and often rapidly receding, whereas those on relatively sheltered coasts are often more subdued.

On the high wave energy coast near Port Campbell in Australia vertical cliffs have been cut by marine erosion in horizontally stratified Miocene calcareous siltstones. The huge waves that break against these during storms have cut out ledges along the bedding-planes at various levels up to 60 m above high tide (Baker 1958). These structural ledges are rarely more than 5 m wide, and are the product of present-day storm wave erosion; they are not emerged shore platforms. The power of storm waves is illustrated where large boulders have been thrown up and over cliffs, as at Quobba in Western Australia.

In NW Australia storm waves produced by tropical cyclones have cut cliff-top features, as at Cape Leveque and Gantheaume Point, near Broome, where a cliff in red Cretaceous sandstone (Broome Formation) rises to a broad bench backed by an upper bluff, which is occasionally trimmed back by storm surge wave overwash in cyclones (Figs. 2.31 and 2.32).

Fig. 2.31 Cape Leveque on the Dampier Peninsula, Western Australia. © Geostudies

Fig. 2.32 Gantheaume Point, Broome, Western Australia. © Geostudies

Cliffs on sheltered sections of the coast, where strong wave action is intercepted by headlands, islands, or offshore reefs, or attenuated by a gently sloping nearshore sea floor, may show profiles partly formed by subaerial weathering and erosion. Cliffs in these situations may have slope-over-wall profiles (§2.4).

Variations in cliff profiles with exposure are well illustrated around Sydney Harbour in New South Wales. The bold cliffs cut in Triassic Hawkesbury Sandstone at the entrance (Fig. 2.33) pass into lower and gentler cliffs and bluffs on the harbour shores, with variations related to the local fetch, which limits the strength of attack by wind-generated waves. The cliffs of Middle Head, facing the entrance from the Tasman Sea and receiving ocean swell and storm waves, are bordered by scrub-covered bluffs on the more sheltered sectors (Fig. 2.34).

Fig. 2.33 Vertical cliffs cut in Hawkesbury Sandstone on North Head, at the entrance to Sydney Harbour. © Geostudies

Fig. 2.34 The cliff at Middle Head interrupts the scrubby bluffs on the coast of Sydney Harbour. © Geostudies

2.3 Bluffs and Steep Coastal Slopes

Bluffs are relatively stable vegetated slopes, usually abandoned cliffs that have been subaerially degraded after marine erosion was halted by the accumulation of a wide protective beach, by emergence due to land uplift or a fall in sea level, by the building of a sea wall or nearshore breakwater, or by land reclamation. Bluffs have slope gradients determined by the physical properties of the rock outcrop: they are usually between 8° and 10° on soft clays and steeper on more resistant formations such as Chalk. Early stages in cliff degradation can be seen on the Chalk coast at Samphire Hoe, near Dover in Kent (Fig. 2.35), where rubble from the Channel Tunnel was deposited on the shore as a fronting terrace between 1987 and 1992.

At Matata on the North Island of New Zealand cliffs became vegetated bluffs after the deposition of a wide protective beach (Fig. 2.36), and emergence due to

Fig. 2.35 The terrace fronting degraded Chalk cliffs at Samphire Hoe, near Dover in Kent. © Geostudies

Fig. 2.36 Bluffs at Matata on the north coast of North Island, New Zealand. © Geostudies

land uplift in the Wellington district in New Zealand during the 1855 earthquake was followed by degradation of cliffs to bluffs. Former cliffs have been subaerially degraded in Denmark because the lowering of the Early Holocene 'Litorina Sea' has withdrawn wave action, the land having risen as the result of postglacial isostatic land uplift to form the marine foreland (Fig. 2.37). Coastal bluffs may also form where cliffs become degraded following artificial protection by sea wall construction, as at Byobuguara on the coast of Japan (Fig. 2.38) or land reclamation. Accretion within Newhaven Harbour in Sussex after the completion of the breakwater in 1891 has led to degradation of the backing Chalk cliff (Fig. 2.39).

Fig. 2.37 The coastal bluff at Mullbjerge in Denmark, fronted by a lowland (the marine foreland) that emerged during the fall of the 'Litorina Sea', about 4000 years ago. © Geostudies

Fig. 2.38 Cliff recession halted by a shore wall at Byobuguara on the east coast of Japan. © Geostudies

Fig. 2.39 The Chalk cliff behind Newhaven harbour in Sussex has been modified following harbour construction. A slope has formed on the capping of Eocene clays, above an apron of downwashed sediment. The formerly vertical cliff is thus developing a convex-over-concave profile. © Geostudies

Fig. 2.40 Steep bluffs near Lynmouth on the North Devon coast. © Geostudies

Fig. 2.41 Hogsback cliff, Trentishoe Down, North Devon. © Geostudies

Bluffs that descend steeply to the shore are seen on the North Devon coast bordering Exmoor (Fig. 2.40). They carry grassland or woodland, and the lower slopes may be actively cliffed. Hogsback profiles occur where the geological strike is parallel to the coast, forming coastal escarpments where the dip is landward (Fig. 2.41).

Steep coastal slopes with a forest cover are extensive in humid tropical regions, where they may show little if any basal cliffing. These profiles result from the intensive subaerial weathering characteristic of the humid tropics, so that many coastal rock formations have been weakened as the result of rapid and deep decomposition by chemical weathering. They form steep coastal slopes with a soil and vegetation mantle, as on the Macalister Range in north Queensland (Fig. 2.42). Wave energy is generally low on humid tropical coasts, many of which are protected from erosion by fringing or nearshore coral reefs. However, these steep forested coastal slopes on deeply-weathered rock formations are subject to occasional landslides, and so to gradual and intermittent coastline recession.

Cliffed coasts are comparatively rare on humid tropical coasts in SE Asia and India, on the east and west coasts of tropical Africa, in Brazil and on the Pacific

Fig. 2.42 Humid tropical forested bluff, Macalister Range, Queensland, on a coast that is protected from strong wave action by the Great Barrier Reef. © Geostudies

coast of Central America, as well as in NE Australia, because of generally low wave energy. On these coasts cliffs occur on exposed promontories, where wave action is relatively strong, and are more extensive on coasts receiving ocean swell, especially where the coastal rock formations are weak or deeply weathered, as in Paraiba, NE Brazil, where 21 % of the coastline consists of retreating cliffs cut in sandy clays (Guilcher 1985). Cliffs shaped by ocean swell are seen along the southern coasts of Indonesian islands, including Bali and Lombok, which receive swell from the Indian Ocean. Yampi Sound in northern Australia is bordered by low crumbling cliffs of deeply-weathered metamorphic rock, from which protrude bolder promontories of quartzite, a type of resistant rock that has been little modified by chemical weathering. Within the humid tropics marine erosion is more effective on coastal rock formations weakened by prior weathering than on their original lithology. On the Liberian coast at Mamba Point a dolerite cliff persists on a headland between vegetated bluffs on outcrops of more thoroughly weathered granite and gneiss.

On arid and semi-arid coasts bluffs may be sparsely-vegetated, as at Blanche Point, south of Port Noarlunga on the coast of South Australia (Fig. 2.43).

On the NE coast of Port Phillip Bay, Australia, bluffs of Tertiary sandstone became unstable when vegetation was cleared from the cliff-top area. This is because the cliff-top vegetation canopy had intercepted some rainfall, and its root systems drew upon water percolating down through the soil, as well as binding surface sediment. When the cliff-top vegetation was cleared runoff and seepage increased, causing erosion of the bluff (Fig. 2.44). Coastal bluffs can be rejuvenated as cliffs when marine erosion is intensified by coastal submergence, or the loss of a protective beach.

Coastal bluffs can be rejuvenated as cliffs when marine erosion is intensified by coastal submergence, or the loss of a protective beach. Bluffs bearing scrub and woodland on the NE coast of Port Phillip Bay have been undercut by wave action following the erosion of beaches that previously protected them (Fig. 2.45). At Sandringham this has been countered by beach renourishment, restoring the protective beach and halting erosion of the backing bluff, which has become revegetated (Fig. 2.46).

Fig. 2.43 A sparsely-vegetated bluff at Blanche Point, South Australia. © Geostudies

Fig. 2.44 Seepage has contributed to slumping on the cliff at Black Rock, Victoria, Australia. © Geostudies

Fig. 2.45 Erosion at the base of the bluff at Sandringham, Victoria, Australia, has followed depletion of the bordering beach. © Geostudies

Some vegetated bluffs show intermittent and local slumping, and could be termed wasting bluffs. Examples are seen on the Pacific coasts of Oregon and Washington, where steep bluffs carrying scrub and forest appear fairly stable, but are in fact receding as the result of intermittent localised slumping (Fig. 2.47).

Fig. 2.46 Beach renourishment has halted bluff-base erosion at the southern end of Sandringham Beach, Victoria, Australia. © Geostudies

Fig. 2.47 Cliffs are formed by occasional slumping on the forested bluffs near Moclips on the coast of Washington State. © Geostudies

Landslides generally occur here in the wetter winter months, during or after occasional storm surges, as the result of earthquakes or in response to a long-term (e.g. 300 year) tsunami cycle (Komar and Shih 1993). The slump scars are quickly revegetated in this cool and moist environment and the debris fans of downwashed or slumped sediment are soon consumed by the sea. Similar processes are effective on vegetated bluffs on humid tropical coasts, such as the Ivory Coast in West Africa.

2.4 Slope-Over-Wall Coasts

Some steep coasts have a composite profile, with an upper slope descending to a basal cliff. Such slope-over-wall coasts can form in several ways. The simplest are on soft homogeneous rock formations where a coastal slope formed by subaerial processes (runoff, slumping, soil creep) descends to a basal cliff kept fresh by wave attack during occasional storms. Such coasts may show alternations in profile, the upper slope being extended by subaerial weathering and runoff during occasional downpours and undercut and steepened to revive the basal cliff after dispersal of downwashed or slumped fans by wave action.

Slope-over-wall profiles are also seen where the upper part of a cliff is a slope cut in weak or weathered rock while the lower part is a vertical cliff in more coherent rock. On the Dorset coast at Portland Bill a coastal slope in the soft Purbeck Beds descends to a cliff cut in harder Portland Limestone (Fig. 2.48), while on the limestone cliffs of south Pembrokeshire thinly-bedded horizontal strata have weathered into a convex upper slope above a vertical cliff of underlying more massive limestone. On Flamborough Head in Yorkshire a grassy upper slope (gradient 20°–30°) cut in soft glacial drift deposits declines to a vertical cliff cut in firmer Chalk, with a basal talus apron (Fig. 2.49).

Fig. 2.48 Slope-over-wall coast at Portland Bill, Dorset. © Geostudies

Fig. 2.49 Slope-over-wall coast at North Landing, Flamborough Head, Yorkshire. © Geostudies

Where a coastal slope follows a bedding-plane in seaward-dipping strata, or seaward-sloping cleavage, joint or fault planes, undercutting by a cliff produces a slope-over-wall profile.

Slope-over-wall profiles have also formed on coasts in high latitudes where cliffs cut in relatively resistant rock are degraded by periglacial freeze-and-thaw resulting in solifluction, forming coastal slopes that are then undercut by marine erosion. This process is still active on Arctic coasts, Baffin Island and Svalbard. It was more extensive during Pleistocene times, when coasts that are now temperate were subject to the down-slope movement of frost-shattered rubble during cold phases when sea level was lower. The Pleistocene cliffs then became slopes mantled by an earthy solifluction deposit with angular gravel (known as a Head deposit) that extended out on to what is now the sea floor in a broad, diminishing apron. Late in Pleistocene times the climate became milder, and these coastal slopes became vegetated. Sea level rose, and marine erosion undercut the Head-mantled slopes. Coastal landforms of this type are found on the west and SW coasts of Britain, where vegetated slopes (typically 20°–30° but locally up to 45°) descend to steeper, rugged, rocky cliffs (Fig. 2.50).

The coastal slope may be a bevel of almost uniform gradient (especially where it follows seaward-dipping bedding, cleavage, joint or fault planes), but more often it is convex in form, like a hog's back, and occasionally it is concave, where the lower slopes of the Head deposit are preserved. Cliffs cut entirely in Head deposits are seen on the coast of Cornwall near St Ives, fronted by platforms in the underlying metamorphic rocks (Fig. 2.51).

The proportion of relict periglaciated slope to actively-receding cliff depends on the degree of exposure of the coast to wave attack. On sheltered sectors of the south coast of Cornwall (as on the eastern side of the Lizard Peninsula near Coverack) the relict slope is well preserved down almost to high tide level. On the more exposed north coast of Cornwall, open to Atlantic storm waves, the relict slope has been undercut by a receding cliff, as on Beeny Cliff, near Boscastle (Fig. 2.52). With

Fig. 2.50 Brown seaward-dipping periglacial Head deposits above a cave in the cliff at Nelly's Cove, Porthallow, Cornwall. © Geostudies

Fig. 2.51 Cliffs cut in Pleistocene Head deposits on the coast at Godrevy, near St Ives in north Cornwall. © Geostudies

Fig. 2.52 Slope-over-wall coast at Beeny Cliff near Boscastle, Cornwall. © Geostudies

even greater exposure to storm wave attack the relict slope has been largely, and in places completely destroyed by Holocene marine erosion, so that little if any of the coastal slope remains on the coast of Watergate Bay in north Cornwall, which has high, almost vertical receding cliffs (Fig. 2.53). The periglaciated coastal slope

Fig. 2.53 An air view of Watergate Bay, Cornwall, where the receding cliffs retain no remnants of the periglaciated slope seen on less exposed parts of the Cornish coast. © Geostudies

is better preserved on the sheltered eastern sides of major headlands, as on Start Point in Devon and Dodman Point in Cornwall, than on their more exposed western shores. Recession of the basal cliff has resulted in increased instability on the upper slope, with rock debris spilling down to the shore.

2.5 Hinterland Effects

Cliffs and bluffs are also influenced by the geomorphology of the immediate hinterland, in particular the topography intersected as they recede. Flat-topped cliffs have been cut into a coastal plateau on the Atlantic coast of Cornwall, particularly south of Portreath, where the even-crested cliffs of Reskajeage Downs rise about 100 m above sea level. Where the hinterland is undulating cliff recession leads to interfluvial ridges becoming promontories between valley-mouth embayments as on the Dorset coast east of Weymouth. This is illustrated on the Yorkshire coast north of Flamborough Head, where cliffs have been cut into a landscape of plateaux and incised valleys on Jurassic formations, forming headlands such as Filey Brigg and embayments such as Robin Hood's Bay. Where the hinterland slopes away from the coast cliffs diminish in altitude as they recede, as at Beachy Head in Sussex and in the vicinity of Childers Cove near Warrnambool in Australia.

References

Baker G (1958) Stripped zones at cliff edges along a high wave energy coast. Proc R Soc Victoria 70:175–179

Bird ECF (1995) Geology and Scenery of Dorset. Ex Libris, Bradford on Avon

Bird ECF (1997) The Shaping of the Isle of Wight. Ex Libris, Bradford on Avon

Bird ECF (1998) The Coasts of Cornwall. Alexander, Fowey

Clayton KM, Shamoon N (1998) A new approach to the relief of Great Britain, II. A classification of rocks based on relative resistance. Geomorphology 25:155–171

Guilcher A (1985) Retreating cliffs in the humid tropics: an example from Paraiba, north-eastern Brazil. Zeitschrift für Geomorphologie, Supplementband 57:95–103

Komar PD, Shih SM (1993) Cliff erosion along the Orgeon coast: a tectonic-sea level imprint plus local controls by beach processes. J Coastal Res 9:747–765

Swale Borough Council (2011) North Sheppey erosion study. www.swale.gov.uk

Chapter 3
Processes on Cliffs

Abstract The main cause of cliff recession is usually wave erosion, but other processes are at work. Undercutting of the base of a cliff by wave scour results in the collapse of rock from the cliff, but instability is also influenced by subaerial weathering, notably the effects of wind, rain and sea spray, seepage of groundwater, recurrent wetting and drying, freezing and thawing, expansion and contraction of the rock surface, the dissolving of soluble rock, and the effects of flora and fauna. Some processes result in the hardening (induration) of rock outcrops in cliffs. Human impacts include coastal mining and quarrying, the addition of buildings, roads and other structures, the dumping of waste and disturbance by traffic, earthworks and explosives.

3.1 Coastal Landslides and Rock Falls

Coastal landslides (also called landslips) are movements of large masses of rock, earth or debris down a coastal slope. Mass movements of this kind occur on weak and weathered rock outcrops and in unconsolidated sediment on steep and cliffed coasts. Instability may be due to the weight of a massive caprock (such as Chalk over sand and clay) and develops with an increase in shear stress (e.g. by greater loading) or a decrease in shear strength (i.e. weakening of the rock formation) exceeds a threshold level and is relieved by movement down-slope (Fig. 3.1). Such movements can take place in various ways (Brunsden and Prior 1984). These include rock falls and toppling, where masses of rock collapse from the cliff face (Fig. 3.2), translational slides, where rock masses slip down a seaward-dipping plane, rotational slides which collapse seaward down a curved plane to form back-tilted rock masses, mudslides where masses of coherent silty or sandy clay move irregularly down across a sheared surface, and mudflows where highly lubricated fine-grained sediment moves downslope as a slurry.

© The Author(s) 2016
E. Bird, *Coastal Cliffs: Morphology and Management*,
SpringerBriefs in Earth Sciences, DOI 10.1007/978-3-319-29084-3_3

Fig. 3.1 Irregular topography produced by mass movement on the coast at White Nothe, Ringstead, Dorset. © Geostudies

Fig. 3.2 Toppling column on cliff of Miocene siltstone near London Bridge, Victoria, Australia. © Geostudies

Coastal landslides are common where geological formations dip seaward, facilitating the sliding of undercut rock down an inclined bedding-plane into the sea, leaving the exhumed bedding-plane as a slope. Recession of a cliff crest also occurs on irregular breakaways, when lumps of rock become dislodged and fall away from the cliff. On the Dorset coast slumping has occurred where cliff-top

Fig. 3.3 Cliff-top breakaway near Charmouth, Dorset. © Geostudies

footpaths have become breakaways (Fig. 3.3). Cracks develop behind, and parallel to, receding cliffs as a prelude to calving or slumping, and deformation or fracturing parallel to the coastline is seen on many cliff-top roads and in damaged buildings close to cliff edges. Little is left of coastal churches at Dunwich in East Anglia and Trzesacz in Poland, both damaged by recession of cliffs cut in glacial drift.

As a rule rock falls leave a scar of paler, unweathered rock on the cliff face, and the debris fan may flow out for a distance at least 4 times the cliff height. Boulder arcs may mark the outer limits of a rock fall several decades later, the finer pebbles and sand having been carried away alongshore and deposited as beaches.

The high cliffs on the west coast of Hawke Bay north of Napier in New Zealand were shaken by earthquakes in 1931 and 2010, causing landslides. A vegetated talus apron is being undercut by the sea to form low earth cliffs (Fig. 3.4).

Fig. 3.4 Grassy talus apron on Moeangiangi cliff in Hawke Bay, New Zealand, resulting from successive earthquakes. © Geostudies

3.2 Effects of Wind Action

Winds blowing against a cliff can winnow fine-grained particles from a rock outcrop, scouring hollows and clefts, and eventually caves (Fig. 3.5), thereby contributing to cliff recession. Displaced sand, silt or clay particles may fall to the shore, or be swept up and over the cliff crest, a process that has formed cliff-top levees on vertical cliffs of Cretaceous calcareous siltstone (calcilutite) on the Port Campbell coast in Victoria, Australia (Fig. 3.6).

Rocky cliffs in Greenland and Antarctica show scoured features resulting from etching by strong winds, especially where they mobilise sandy material.

Fig. 3.5 Cliff cave cut in Pleistocene dune calcarenite Jubilee Point, Nepean coast, Victoria, Australia. © Geostudies

Fig. 3.6 Cliff-top levee near Port Campbell, Victoria, Australia. © Geostudies

3.3 Effects of Runoff

Rain and melting snow can generate water flowing down a cliff face, which on soft rock outcrops washes away sediment, forming rills that may grow into gulleys (Fig. 3.7). Downwashed sediment forms fans at the bottom of gullies, and aprons along the base of the cliff (Fig. 3.8). Generally these are swept away by wave action, exposing the cliff base to further wave attack, but if they persist and grow the cliff eventually declines to a depositional slope. Cliff-face rills and gulleys do not persist on cliffs that are being cut back rapidly by marine erosion.

Cliffs cut in Lower Cretaceous (Wealden) clays and sands on the south-west coast of the Isle of Wight have been dissected by gulleys cut by runoff after heavy rain. A similar cliff of soft Tertiary sandstone at Black Rock Point, on the NE coast of Port Phillip Bay, Australia, has been dissected by vertical gulleys cut by occasional runoff (waterfalls poured over the cliff here during a heavy downpour in 1972), accompanied by the exudation of fine sediment by groundwater seepage. This cliff has receded as the result of the incision and integration of these gulleys (Fig. 3.9). It became steeper and smoother in profile after a cliff-top drain was inserted to divert

Fig. 3.7 Gulleys and rills on cliff at Torrey Canyon, California. © Geostudies

Fig. 3.8 Cliff rills and downwashed cliff-base aprons, Black Rock Point, Victoria, Australia. © Geostudies

runoff (Bird and Rosengren 1987) (Fig. 3.10). Subaerial erosion (rilling) continues here even when the cliff base is not being cut back by marine erosion.

On the north coast of France the Chalk cliffs at Le Petit Ailly have been stained by downwash from a capping of Tertiary sand and clay (Fig. 3.11).

3.4 Effects of Sea Spray

Sea spray can also generate runoff down a cliff. It also contributes to weathering as it dries, when crystallising salt may pluck the rock surface, forming pits that grow into honeycombs (Fig. 3.12). Some limestone coasts have cliffs pitted by solution in rainwater and sea spray, while at lower levels they have been smoothed by abrasion, where waves armed with sand and gravel have scoured the rock surface. Commonly rock surfaces above high tide level are whitened by the deposition of a thin covering of salt during dry periods, and this disappears when the rocks are splashed by rain or spray (Fig. 3.13).

Rocky coasts around the Baltic Sea are weathered by freezing sea spray.

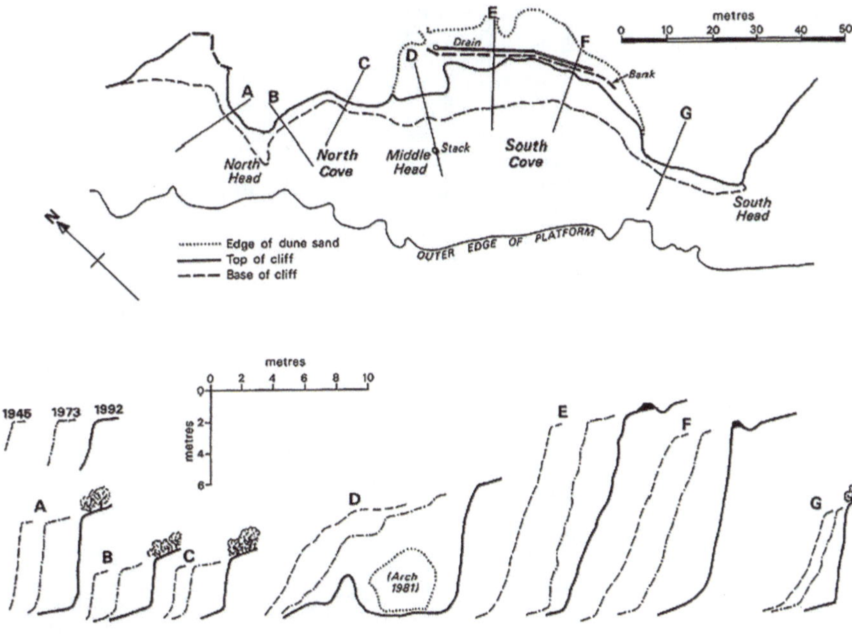

Fig. 3.9 Cliff recession at Black Rock Point, Victoria, Australia. © Geostudies

Fig. 3.10 Cliff-top drain inserted to intercept runoff at Black Rock Point, Victoria, Australia. © Geostudies

3.5 Effects of Groundwater Seepage

Apart from runoff, the water produced by rainfall or melting snow percolates as groundwater into rock that is permeable because of cavities, fractures or interstitial spaces (Fig. 3.14). Water soaks into fine-grained limestone or Chalk. On cliffs cut in glacial drift near Port Angeles, Washington State, it was observed that seepage increased when cliff-top lawns and gardens were watered.

Fig. 3.11 Downwash on cliff at Le Petit Ailly, Normandy, France. © Geostudies

Fig. 3.12 Honeycombed cliff, Otways coast, Victoria, Australia. © Geostudies

Fig. 3.13 Salt precipitated on sandstone, Quiet Corner, Black Rock, Victoria, Australia. © Geostudies

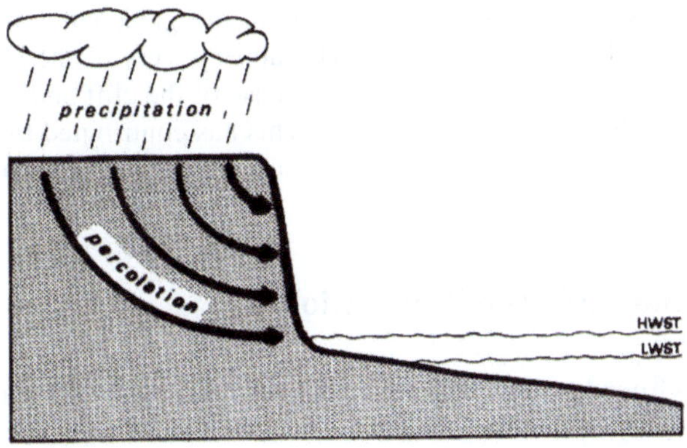

Fig. 3.14 Groundwater seepage and cliff exudations. © Geostudies

Fig. 3.15 Coastal landslide at Folkestone Warren, Kent. © Geostudies

Groundwater seepage from a cliff face can wash out fine-grained sediment, leaving cracks and crevices, and accumulating cliff-foot talus aprons. Surface exudations of fine sand may become cemented to cliff faces on sandstones by precipitation of carbonates, forming vertical ridges. Strong seepage produces runoff, excavating rills and gulleys and generating downwash. Seeping groundwater can also dissolve soluble rock, notably carbonates, etching cracks and cavities in limestones. When the rock in a cliff or bluff becomes saturated with groundwater slumping may occur, particularly on outcrops of clay, siltstone or soft sandstone.

On the south coast of England there have been major coastal landslides where permeable Chalk and Upper Greensand formations dip seaward and rest on impermeable Gault. Water seeping through the permeable rocks flows down the inclined clay surface, and if marine erosion has exposed the junction at or above sea level this lubrication of the interface leads to slipping of the overlying rocks. A spectacular landslide occurred in this situation on the Devon and Dorset coast between Axmouth and Lyme Regis in 1839 (Bird 1995).

The physical effects of wetting and lubrication of the clay surface can be accompanied by a chemical process, for when seeping water rich in dissolved calcium carbonate reaches the glauconitic Gault clay base exchange occurs, calcium

ions displacing potassium ions in the glauconite so that alkalinity increases and the clay is deflocculated. The upper layers of the Gault clay then become a soft wet slurry that flows out at the base of the cliff and hastens the undermining of the Chalk and Upper Greensand. This has contributed to slumping where the rock formations dip seaward, as at Folkestone Warren in Kent (Fig. 3.15).

3.6 Effects of Saturation

Groundwater accumulation in permeable rock formations can contribute to instability. When they become saturated the additional loading increases shear stress, which explains why many coastal landslides are most active during wet periods. Groundwater loading of porous coastal Head formations (unconsolidated earthy rubble formed by past periglaciation) has led to cliff collapses at Downderry and Gunwalloe on the south coast of Cornwall. Groundwater saturation and seepage contributed to cliff instability at Church Cliff, Lyme Regis, where drainage systems have been introduced to reduce groundwater pressure (Fig. 3.16). In December 2000 after a wet autumn saturated clay and rubble spilled over the cliffs and down across the beach at Black Ven, east of Lyme Regis.

3.7 Effects of Wetting and Drying

Recurrent wetting by rain or dew and subsequent drying results in disintegration of rock outcrops on cliffs, particularly on fine-grained rock, while drying of saline sea spray adds the plucking effects of salt crystallisation.

Some clays (smectites) swell in volume as they become wet because of water absorption, and shrink again as they dry out. Such expansion and contraction can cause instability, particularly on clay slopes, leading to slumping and landslides.

Fig. 3.16 Drainage pipes inserted in the 1980s to evacuate groundwater from Church Cliff at Lyme Regis, Dorset. © Geostudies

Fig. 3.17 Desiccation caused weathering of clay on cliffs at Kimmeridge in Dorset during a dry summer. © Geostudies

Montmorillonite is the commonest of the smectite clay minerals, while kaolinite is less susceptible to these changes, and so more stable.

By contrast, prolonged desiccation can lead to flaking and falling of particles from cliffs. In the dry English summer of 1976 basal fans of collapsed clay particles formed beneath cliffs cut in a Jurassic formation at Kimmeridge on the Dorset coast (Fig. 3.17). A 30 m long sector of the cliff of Triassic sandstone east of Sidmouth in Devon collapsed in June 2001 after a prolonged spell of dry weather, when shrinkage of the rock may have been due to desiccation.

3.8 Effects of Thermal Expansion and Contraction

Changes in temperature (particularly freezing and thawing) lead to expansion and contraction of rock outcrops and their disintegration (Fig. 3.18). These processes are responsible for fracturing, flaking and spalling on cliff faces, particularly along zones of weakness such as joints or bedding-planes.

Fig. 3.18 Chalk shattered by Pleistocene freeze-and-thaw action exposed in a cliff at Watcombe Bay, Isle of Wight. © Geostudies

Slumping may also occur as a consequence of changes in rock volume as ground temperatures rise and fall. Many rock falls occur on cliffs during unusually cold weather, especially when stresses develop as the result of freezing and thawing of groundwater contained in the rock formation. Rock falls on Chalk cliffs in southern England occur particularly during and after cold winters in response to freezing and thawing of groundwater in the Chalk; there is expansion and contraction of ice as it forms and melts. Cliff faces show paler segments of steeply inclined bedding planes where masses of grey weathered Chalk have fallen away, producing a basal apron of rubble that is gradually removed by wave erosion (Fig. 3.19). Incipient slumping is indicated by cracks and services in a cliff (Fig. 3.20).

Rock falls often occur on cliffs at night or in winter, perhaps because of stresses caused by shrinkage as the nocturnal temperature falls. A major collapse occurred on Beachy Head, Sussex, in January 1999 on a 200 m long and 90 m high section of Chalk cliff. It left a white scar on the otherwise grey Chalk cliff and produced a debris fan of up to 5 m thick (volume 50,000 m^2) out to and beyond the lighthouse (Williams et al. 2004).

Reference has been made (§2.4) to periglacial processes active on high latitude coasts, where rocky outcrops are frost-shattered, producing cliff-base talus fans of angular, unsorted debris. At Cape Lisburne in Alaska cliff recession by frost shattering has produced an unvegetated talus slope over 300 m high descending into the sea, and similar features are seen on Baffin Island in northern Canada, and on Svalbard (Klemsdal 2010), in Chilean Patagonia and around Antarctica. Frost shattering has led to rapid cliff recession on the Pacific coast of Sakhalin in eastern Russia (Zenkovich 1967).

On the island of Rügen on the German Baltic coast there are Chalk cliffs, fronted by flint shingle and glacial boulders. Aprons of frost-shattered rubble form each winter, particularly when there is a protective shore fringe of sea ice, and are undercut and washed away by the summer sea (Fig. 3.21).

Fig. 3.19 Chalk rock fall and scar at Alum Bay, Isle of Wight. Slumping of this kind usually occurs during very wet weather, or when a thaw follows freezing. © Geostudies

Fig. 3.20 Crack in cliff of Miocene siltstone on Island Archway, near Port Campbell, Victoria. © Geostudies

Fig. 3.21 Frost-shattered rubble apron on a Chalk cliff on Rügen, Germany. © Geostudies

In northern Alaska and on the arctic coasts of Canada and Russia the summer thawing of frozen ground (termed permafrost) contributes to rapid degeneration of tundra cliffs that consist of peat and morainic sediments. The frozen sediments are coherent, but when they thaw they become incoherent, and as shear strength diminishes the cliffs collapse. The sediments slump into the sea and are swept away by wave action (Selivanov 2010). Seasonal freezing and thawing on tundra cliffs can cause them to retreat tens of metres in a few weeks.

3.9 Effects of Solution

Soluble rock, notably limestone, is etched by solution in rainfall (which contains dissolved carbon dioxide, and is therefore a weak acid) and percolating groundwater (providing it is not already saturated with dissolved carbonates), so that a limestone cliff face becomes pitted and grooved (Fig. 3.22). This kind of weathering is termed karstic. It occurs also on cliff outcrops of dune calcarenite (see below).

3.10 Effects of Plants and Animals

Coastal vegetation and fauna often include species, the growth and metabolism of which lead to decomposition or dissolution of rock outcrops, especially limestones: a process termed bioerosion. Plants growing on or near cliffs can cause erosion by the physical effects of root penetration and expansion, widening joints and fissures (Fig. 3.23); they may also cause chemical weathering by the accompanying effects of corrosive liquids extruded from roots. Pitting of chalk and limestone surfaces by corrosion associated with algal encrustations also contributes to cliff erosion.

Fig. 3.22 Karstic weathering on a dune calcarenite cliff at Sorrento, Victoria, Australia. © Geostudies

Burrowing animals such as rabbits and foxes can cause erosion on cliffs and steep coastal slopes. Nesting holes excavated by birds, particularly sand martins, contribute to the erosion of soft sandstone cliffs on the Suffolk coast (Fig. 3.24), while burrowing mortar bees have dissected the sandy cliff near Redend Point in Dorset (Fig. 3.25).

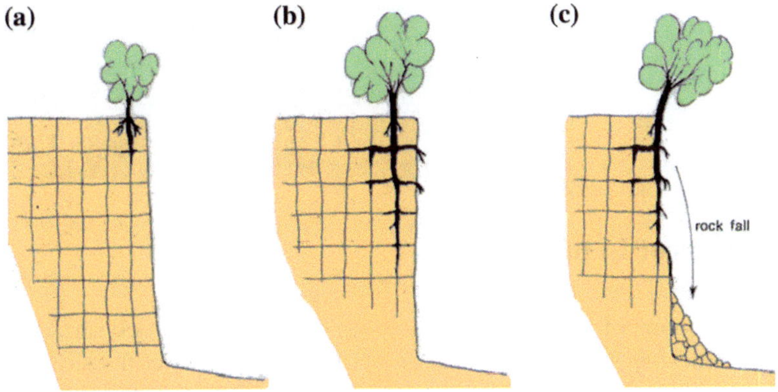

Fig. 3.23 Root penetration widens joints and can cause rock falls. © Geostudies

Fig. 3.24 Sand martin burrows in a sandstone cliff at Minsmere, Suffolk (Bioerosion). © Geostudies

Fig. 3.25 The cliff in Eocene sandstone at Redend Point, Dorset, has been burrowed by bees (Bioerosion). © Geostudies

3.11 Effects of Induration

While many forms of weathering decompose or disintegrate rock surfaces, some result in hardening (induration). Superficial induration of a cliff face can result from the precipitation of cementing materials (such as carbonates or ferruginous or siliceous cementing compounds) washed down the cliff or brought to the surface by groundwater seepage. Such 'case-hardening' impedes erosion, at least until the hardened crust is breached or dissected by erosion. The vertical cliff face at Demons Bluff on the coast of Victoria, Australia, shows slumping and cavitation in sectors where the indurated crust has been breached to expose softer underlying rock weakened by the withdrawal of cementing compounds to the case-hardened surface (Fig. 3.26). The cliffs retreat by successive breakaways where vertical sheets of case-hardened rock separate from the cliff face, then collapse, disintegrating as they fall to the shore, forming a heap of rubble that is gradually dispersed by wave action.

Various kinds of rock can become indurated on cliff faces, making them much harder than in fresh exposures in inland quarries, or in boreholes in the same

Fig. 3.26 Breakaways causing cliff recession on Tertiary silty sandstone at Demons Bluff, Victoria. © Geostudies

formation. In South Devon Beer Stone is a fine-grained homogenous Chalk, moist when freshly cut in underground caverns, and so useful for carving or ornamentation on buildings, notably churches and cathedrals, but it soon dries and hardens on exposure to the atmosphere: it is much softer than the same Chalk outcrop in nearby coastal cliffs.

Some coastal rock outcrops have been modified by weathering and induration: in the humid tropics laterisation can result in the formation of a hardened caprock above a weakened pallid zone: iron oxides leached from the pallid zone have been carried upward in groundwater (explain) and deposited in the overlying caprock. Red Bluff on Port Phillip Bay has a profile related to prior laterisation, including an indurated caprock formed during a late Pliocene episode when the climate was warmer and wetter than it is now (Fig. 3.27).

In Australia Pleistocene dune calcarenite is found where calcareous sand dunes have become lithified by precipitation of calcium carbonate from percolating groundwater. Cliffs cut in this formation show further induration as the calcareous sand formation hardens on exposure to the atmosphere as the result of precipitation of carbonates from seeping groundwater, sea swash or spray. The pale yellow calcareous sand becomes dark grey as induration occurs (possibly associated with colonisation by algae). Weathering by rainfall or aerated sea spray may

Fig. 3.27 Laterised outcrop of Pliocene sandstone at Red Bluff, Melbourne, Victoria. © Geostudies

then produce spiky karstic surfaces (lapiés) on cliff faces (Fig. 3.22). Where the hardened cliff face is breached by erosion or a rock fall the paler underlying soft calcareous sand is exposed, and may be excavated by wind action, forming a cave (Fig. 3.5). The exposed calcareous sand is then gradually hardened by carbonate precipitation from seepage, a process that takes a few decades (Bird 1993). On the Nepean coast in Victoria, cliff faces have been indurated by carbonate precipitation from sea spray to a higher level on headlands exposed to strong wave action than behind intervening more sheltered bays, so that cliff erosion and slumping are more common in the bays.

In a similar way ferruginous sandstones (composed of quartz sand grains coated with ferrous oxides) may become superficially indurated on exposure to the atmosphere. The iron oxides dissolve in percolating groundwater, and as the rocks dry out they are drawn towards the rock surface, where they are concentrated and precipitated within interstitial cavities. The ferrous iron compounds then oxidise to form less soluble ferric minerals, such as brown goethite and reddish haematite, which bind the sand grains more firmly and produce a hardened sandstone (Bird and Green 1992). An example of this is seen in the cliffs of Miocene sandstone on the north-east coast of Port Phillip Bay, where a soft light brown silty sandstone formation seen in inland exposures (and containing the pale grey or green mineral glauconite, a complex silicate of aluminium, magnesium and both ferrous and ferric iron), has been hardened and darkened in cliff faces that have been frequently splashed by the sea (the glauconite converted to goethite) (Fig. 3.28). There is a transition from dark brown, hard sandstone on headlands that are frequently splashed to paler yellow and light brown soft sandstone exposed in receding cliffs in adjacent more sheltered bays. A similar transition runs upward in headland cliffs from hard dark brown rock in the intertidal zone to softer yellow and light brown sandstone above the high tide swash limit. Where rock falls occur there are paler scars in the cliff where softer underlying rock is exposed, and these darken and harden as the iron compounds oxidise over a period of several years.

Fig. 3.28 Induration of Miocene sandstone, Mount Eliza, Victoria, Australia. © Geostudies

3.12 Human Impacts

Cliffs and bluffs can be modified, directly or indirectly, by human impacts such as the quarrying of rock or sand (Fig. 3.29), the making of tracks or stairways, or the building of structures on or against cliff faces (Fig. 3.30). Cliffs of limestone and clay at Ravenscar in Yorkshire were much disturbed historically by quarrying for alum, and examples of cliffed coasts modified by quarrying are seen on Portland Bill in Dorset, at Stepper Point and near Perranporth on the north coast of Cornwall. Many coastal quarries were cut to produce rock for use in harbour breakwaters or shore protection schemes, as at Portland in Victoria, Australia.

Some cliffs have been converted into stable artificial slopes, as in several English seaside resorts, notably at Brighton, where former Chalk cliffs that extended eastward from the Aquarium and Kemp Town are now walled or vegetated bluffs behind a broad Marine Parade. Similar stabilised cliffs are seen at Bournemouth, Ramsgate and Clacton.

Fig. 3.29 Cliff quarry to provide stone for the building of nearby Black Rock House, Melbourne, Victoria, Australia in the 1850s. © Geostudies

Fig. 3.30 Building against a cliff on the Durham coast. © Geostudies

Near Newport in Oregon, the loading of a cliff-top by buildings was followed by landslides that damaged or destroyed many houses and led to the demolition of a condominium (Komar 1983; Viles and Spencer 1995). The weight of cliff-top buildings was partly responsible for the collapse of Holbeck Hall Hotel on a steep coastal slope in glacial boulder clay at Scarborough in Yorkshire in 1993.

The building of roads along steep coasts often results in instability, both during their construction when the slope is excavated and debris spills down to the shore, and when the road is disrupted by slumping or rock falls. This was a problem during the construction of the Great Ocean Road, a corniche road on the steep forested coast of the Otway Ranges in Victoria, Australia between Eastern View and Apollo Bay in the 1920s, and recurrent instability has continued to make this a hazardous highway, expensive to maintain (Fig. 3.31).

An alternative is to build a highway bridge along the coast. The Corniche Road at Coalcliff, south of Sydney in New South Wales proved difficult to maintain because of recurrent cliff falls (Fig. 3.32) and so a road was constructed along Sea Cliff Bridge in 2004 (Fig. 3.33).

North of Teignmouth in Devon cliffs of red sandstone and conglomerate cliffs extend to Dawlish, and the coast was modified by the construction of the Great Western Railway in 1846. The cliff is protected by a sea wall and promenade (Fig. 3.34), and there are tunnels through headlands. The problem of rocks and

Fig. 3.31 The Great Ocean Road, cut into the steep coast of the Otway Ranges, Victoria, Australia. © Geostudies

Fig. 3.32 The former corniche road at Coalcliff, New South Wales, on a vertical cliff of Triassic sandstone. © Geostudies

Fig. 3.33 Sea Cliff Bridge, along the coast at Coalcliff, south of Stanwell Park, New South Wales. © Geostudies

Fig. 3.34 A section of the Great Western Railway runs along the South Devon coast between Teignmouth and Dawlish, in front of sandstone cliffs, which have been protected by fencing and wire mesh to prevent rocks falling on to the railway lines. © Geostudies

earth slumping from the cliff on to the railway has been countered by introducing fencing and covering parts of the cliff with wire mesh.

On the Beaumaris coast in south-eastern Melbourne vibrations caused by heavy lorry traffic on the coast road have disturbed coastal rock formations and contributed to recurrent rock falls in the bordering cliff.

The use of explosives to blast cliff faces, notably where it has been considered necessary to remove dangerous overhanging rock, is likely to increase long-term instability by shattering and weakening the rock formations. Limestone cliffs near Llantwit Major, in South Wales, remained unstable after they were blasted in 1969, and it is possible that the collapse of a cliff cave at Gracelands in Western Australia was a consequence of previous use of explosives on the cliff. It would be wise to refrain from using explosives on coastal cliffs.

Rock falls have occurred from time to time on sectors of the Port Campbell calcareous siltstone cliffs in Victoria, Australia (Fig. 3.35), and disturbances by seismic testing offshore in search of oil and natural gas may have increased the instability of cliffs on this coast.

Fig. 3.35 Collapsed cliff near Port Campbell, Victoria, Australia. © Geostudies

Fig. 3.36 Rubble waste dumped on the cliff at Red Bluff, Victoria, Australia. © Geostudies

Cliffs have been modified by the dumping of waste material, as at Fort Bragg in California, where broken glass, porcelain and other garbage were deposited from the cliff crest, a procedure that was halted in 1970. At Red Bluff, Sandringham on the coast of Port Phillip Bay, Australia rubble waste from roads and buildings was dumped over the cliff for several years. Since the 1960s the waste-mantled slope has been partially concealed by grassy vegetation (Fig. 3.36).

Structures have been built on cliffs to launch hang-gliders at Stanwell Park, New South Wales, and to observe marine animals from the bluff at Sea Lion Point on the Oregon coast.

References

Bird ECF (1993) The coast of Victoria. Melbourne University Press
Bird ECF (1995) Geology and scenery of Dorset. Ex Libris, Bradford on Avon
Bird ECF, Green N (1992) Induration of ferruginous shore rock outcrops. Victorian Naturalist 109:64–69
Bird ECF, Rosengren NJ (1987) Coastal cliff management: an example from Black Rock Point, Melbourne, Australia. J Shoreline Manage 3:39–51
Brunsden D, Prior DB (1984) Slope instability. Wiley, Chichester
Klemsdal T (2010) Svalbard and Jan Mayen. In: Bird ECF (ed) Encyclopedia of the World's Coastal Landforms. Springer, Dordrecht, pp 581–586
Komar PD (1983) Handbook of coastal processes and erosion. CRC Press, Boca Raton, Florida
Selivanov A (2010) The Arctic coast of Russia. In: Bird ECF (ed) Encyclopedia of the World's Coastal Landforms. Springer, Dordrecht, pp 823–829
Viles H, Spencer T (1995) Coastal problems. Arnold, London
Williams RBG, Robinson DA, Dornbusch U, Foote YM, Moses CA, Saddleton PR (2004) A sturzstrom-like cliff fall on the chalk coast of Sussex. In: Mortimore RN, Duperret A (eds) Coastal chalk cliff instability. Engineering Geology Special Publication, vol 20, pp 89–97
Zenkovich VP (1967) Processes of coastal development (trans: Fry OG, ed. Steers JA). Oliver and Boyd, Edinburgh

Chapter 4
Rates of Cliff Recession

Abstract Cliffs recede as the result of basal erosion, slumping, landslides and rock falls, and the removal of collapsed debris from the cliff base. Cliff recession occurs particularly during storms, when large waves attack the cliffs, or during events such as heavy and prolonged rainfall, the melting of snow or frozen ground, or earthquakes. The rate of cliff retreat depends on the frequency and severity of these events, but is also influenced by the resistance and structure of the outcropping formations, tide range and the presence of a rocky shore or shore platform that affords protection to the cliff base. Cliff recession is generally cyclic, with intermittent landslides or rock falls followed by the removal of slumped or downwashed aprons of rock debris or sediment.

Where cliffs expose rocks that are very resistant (e.g. massive granites) coastlines have changed little, if at all, as the result of marine erosion over the past 6000 years. Less resistant formations have been cut back as cliffs, some bordered by irregular rocky shores, others with smoother shore platforms at least partly exposed at low tide. Elsewhere there are cliffs or coastal slopes that plunge into deep water. A distinction is sometimes made between hard rock cliffs and soft rock cliffs, also known as earth cliffs (May 1972), which are often sloping or slumping, and vertical only when they have been trimmed back by basal wave attack, often only temporarily, with a return to a sloping or slumping profile between storm episodes.

Cliffs can recede quickly on earth cliffs or other soft sediments such as alluvium, glacial drift or volcanic ash. Rapidly receding vertical cliffs up to 2 m high have been cut into peat bog deposits in NW Ireland where blanket bogs descend to sea level around Broad Haven and Blacksod Bay. Similar rapidly receding vertical cliffs have been cut in freshwater swamp deposits (humates) at Lang Lang on the north-east coast of Westernport Bay in Victoria, Australia (Fig. 4.1), at Owenga on the east coast of the Chatham Islands and near Budir in NW Iceland.

© The Author(s) 2016

E. Bird, *Coastal Cliffs: Morphology and Management*,
SpringerBriefs in Earth Sciences, DOI 10.1007/978-3-319-29084-3_4

Fig. 4.1 Earth cliff, Lang Lang, Western Port Bay, Victoria, Australia. The cliff is cut into humate, a freshwater swamp deposit. © Geostudies

Fig. 4.2 Sloping cliff in glacial drift on the coast of the Bay of Fundy, Canada. © Geostudies

Wave energy varies with coastal aspect, being stronger on exposed sectors than on those sheltered by peninsulas or islands, or where the nearshore sea floor profile is gently sloping. Where the tide range is small wave energy is concentrated in a narrow vertical zone, and where the tide range is larger wave energy is more dispersed. On the megatidal shores of the Bay of Fundy there are sloping cliffs in soft Pleistocene glacial drift which retreat only during occasional brief episodes when high spring tides bring wave action to the cliff base (Fig. 4.2). Cliff recession is also influenced by the incidence and frequency of rock falls and landslides or cliff falls (when the crest of the cliff may retreat, while the cliff base may advance, at least temporarily, by the accumulation of a lobe of slumped material).

Cliff recession rates diminish on emerging coasts, such as those of the Gulf of Bothnia, where wave erosion has been withdrawing seaward across an emerging sea floor and abandoned cliffs are degraded to bluffs. Evidence that cliff recession has been slow includes the persistence of archaeological structures (such as Nero's Palace on the coast of Anzio in Italy), or the survival of mediaeval fish traps (as at Aberath on the west coast of Wales).

Measurements of cliff recession rates are most readily made at the cliff crest, but it is also possible to measure basal cliff retreat or the recession of the whole profile, the changing gradient depending on the relative rates of retreat of the

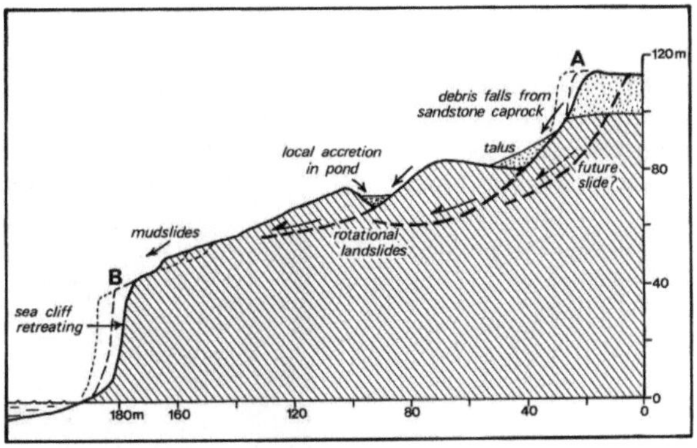

Fig. 4.3 Recurrent landslides on the coast near Lyme Regis in Dorset, showing recession of the cliff crest (A), accompanied by recession of the basal cliff (B). © Geostudies

cliff base and at the cliff crest. There is sometimes a cascade effect, in which the removal of sediment from the cliff base by waves and currents causes instability, which is transmitted upward to the cliff crest (Fig. 4.3), which then begins to recede. Slumping coastal slopes are irregular, and basal debris fans are undercut by the waves, forming a slope-over-wall profile (§2.4). The vertical cliff grows in height as it is cut back, but there is soon further slumping. The cliffs thus recede as the result of alternating marine and subaerial erosion (Brunsden and Jones 1980).

Cliff recession is often instantaneous (as when rock falls or landslides occur), but may take place over time scales ranging up to several centuries. Most existing cliffs have been cut back within the past 6000 years, when the sea has stood at or close to its present level. On some coasts it is possible to measure the extent of this recession with reference to remains of the preceding subaerially weathered land surface (oxidised rock) intersecting the nearshore sea floor. Gill (1973) used this method to show that the cliffs on the Otway coast in Australia had receded 105 m on mudstones and 53 m on sandstones over the past 6000 years. It may be possible to reconstruct the profile before a cliff was cut, as on the flanks of a coastal or island volcano, or on a slope-over-wall coast, and so estimate the extent and rate of recession. On the north coast of Cornwall where islets and stacks such as those at Bedruthan Steps and Tobban Horse near Porthtowan retain segments of a periglaciated rubble-drift (Head) slope it is possible to estimate the extent of cliff-base recession during the last 6000 years, since the sea rose to its present level: typically at least 100, and locally as much as 200 m (Bird 1998) (Fig. 4.4).

Rates of cliff-crest recession can be measured by repeated surveys, by photogrammetric analysis, by measurements from inserted pegs or using micro-erosion meters. Cliff recession is usually expressed as an average in metres per year, but the actual retreat of a cliff crest is episodic and localised as each rock fall occurs. Measurements of average rates of cliff recession range from up to a millimetre/year on cliffs cut in granite to up to a centimetre/year on limestone or shale, up to a metre/year on chalk and sandstone and several metres/year on glacial drift deposits or volcanic ash (Sunamura 1992).

Cliff-crest recession in glacial drift on parts of the Polish coast has averaged a metre a year, with up to 5 m in a stormy year. May and Heeps (1985) found rates of cliff-top recession ranging up to more than 1 m/year at various sites on

Fig. 4.4 Cliff recession on the coast near Porthtowan, north Cornwall, indicated by the survival of a segment of periglaciated slope on the stack at Tobban Horse. The pecked line indicates the restored slope profile. © Geostudies

Chalk coasts. On Chalk cliffs in Sussex an area of 3264 m^2 of coastal land was lost over 12 years, almost all of which was removed in winter periods, and 42 % in two particularly cold, wet and stormy winters by way of intermittent and localised rock falls.

Comparison of the Chalk coastline of South Foreland in Kent from maps dated 1878 and 1962 showed an average annual retreat of 0.19 m (May 1971). This suggests that the coastline here has retreated nearly 400 metres since Julius Caesar passed this way in 55 BC on his way to invade England, while an average retreat of the Chalk cliffs bordering the Strait of Dover at this rate would have set them back more than a kilometre since the sea rose to its present level about 6000 years ago.

Because of the intermittency of slumping average annual rates of cliff retreat can be misleading. Along the Port Campbell cliffs in Australia the mean rate of cliff recession is only a few centimetres per year, achieved by a series of occasional localised rock falls, when segments up to 200 m long and 12 m wide have suddenly collapsed into the sea (Baker 1943). One such collapse in 1939 near Sentinel Rock left a freshly exposed scar that is still discernible as a less-weathered cliff sector 75 years later, while a basal apron of fallen debris is being slowly consumed by marine erosion (Fig. 4.5). Evidence of such slumping is present on

Fig. 4.5 Segment of scar (indicated by X) left by a 1939 rock fall on Sentinel Rock cliff, Port Campbell, Victoria, Australia. © Geostudies

about 5 % of the length of the Port Campbell cliffs, suggesting a recurrence interval of many decades for such major events.

The cliff crest and cliff base at Scarborough Bluffs on the north shore of Lake Ontario have both retreated at an average rate of up to 0.5 m/year, mainly during episodes of storm wave attack in high lake level phases and as the result of recurrent slumping (Carter and Guy 1988). Repeated surveys on North Shore Bluffs, on the Ontario coast of Lake Erie in Canada by Quigley and Di Nardo (1980) found cliff-crest retreat of 2.2 m/year and cliff-base retreat of 1.5 metres per year between 1964 and 1977, accompanied by variations in slope profile, as shown in Fig. 4.6.

Where sea walls have been built to halt cliff recession, particularly at seaside resorts, a protected cliff sector may now stand forward from the coastline because of continuing cliff recession on the adjacent unprotected coast. Examples of this are seen on glacial drift coasts at Withernsea and Mundesley in eastern England, at Ustronie Morskie on the Polish coast, and at Kampen on the German North Sea island of Sylt, where a short sector protected by artificial structures soon projected seaward from the receding cliffs on either side.

Accumulation of large quantities of fallen rock on the shore protects the base of a cliff from wave attack because storm wave energy is expended on the fallen rock. Deposition of a wide protective beach can also reduce or halt cliff recession. Small quantities of beach material that can be mobilised and hurled on the cliff base during storms accelerate abrasion by waves. On the Chalk cliffs of Normandy in France a wide flint shingle beach is protective, but where it narrows the gravel is thrown against the cliff base, undermining it and accelerating cliff retreat (Costa et al. 2006).

On the Holderness coast in eastern England measurements of cliff erosion have shown local acceleration with the passage of low sectors of the southward-drifting beach (Pringle 1981). Cliff recession on Seatown Beach, Dorset, has been more rapid near the western end, where the shingle beach is lower and finer than it is to the east.

An unusual form of cliff recession occurred at Fairhaven, Washington State, late in the 19th century, when hydraulic sluicing was used to wash sand and gravel from Post Hill Point, a bluff of glacial drift, to provide material to fill and reclaim Commercial Point in an adjacent bay. The cliff is now stable and vegetated behind

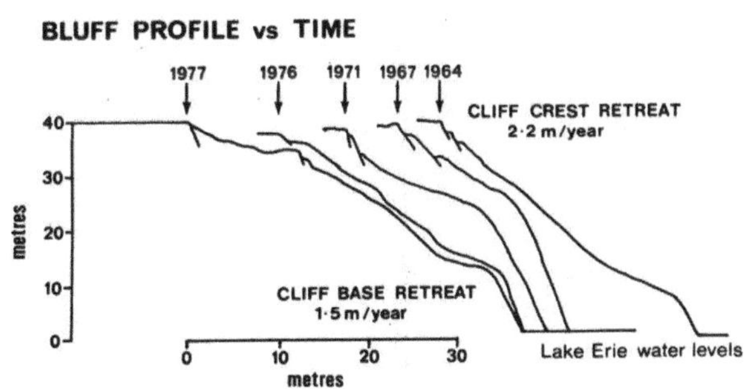

Fig. 4.6 Variations in slope profile on a retreating clay cliff. © Geostudies

a reclaimed area of shipbuilding yards, a ferry terminal and factories. Similar sluicing occurred on nearby Whidbey Island (Schwartz and Terich 2010).

Demands for stabilisation of cliffs in their present position should therefore bear in mind that those cut in rocks of moderate resistance, such as sandstone or chalk, have receded considerable distances during the past few thousand years, and that such recession will continue in the future as the result of coastal processes. Reference has also been made to the fact that retreating cliffs are a source of sediment supply to the shore, nourishing nearby beaches (§3.1).

References

Baker G (1943) Features of a Victorian limestone coastline. J Geol 51:359–386

Bird ECF (1998) The coasts of Cornwall. Alexander, Fowey

Brunsden D, Jones DKC (1980) Relative time scales in formative events in coastal landslide systems. Z Geomorphol, Supplementband 34:1–19

Carter CH, Guy DE (1988) Coastal erosion: processes, timing and magnitude at the bluff toe. Mar Geol 84:117–130

Costa S, Hénaff A, Lageat Y (2006) The gravel beaches of north-west France and their contribution to the dynamic of the coastal cliff-shore platform system. Z Geomorphol, Supplementband 144:199–214

Gill ED (1973) Rate and mode of retrogradation on rocky coasts in Victoria, Australia. Boreas 2:143–171

May VJ (1971) The retreat of chalk cliffs. Geogr J 137:203–206

May VJ (1972) Earth cliffs. In: Barnes RSK (ed) The Coastline. Wiley, New York, pp 215–235

May VJ, Heeps C (1985) The nature and rates of change on chalk coastlines. Z Geomorphol, Supplementband 57:81–94

Pringle A (1981) Beach development and coastal erosion in Holderness. In: Neale J, Flenley J (eds) The quaternary in Britain. Pergamon, Oxford, pp 194–205

Quigley RM, Di Nardo LR (1980) Cyclic instability modes of eroding clay bluffs, Lake Erie Northshore Bluffs at Point Bruce, Ontario, Canada. Z Geomorphol, Supplementband 34:39–47

Schwartz ML, Terich T (2010) Washington. In: Bird ECF (ed) Encyclopedia of the World's Coastal Landforms. Springer, Dordrecht, pp 15–31

Sunamura T (1992) Geomorphology of rocky coasts. Wiley, Chichester

Chapter 5
Cliff Hazards

Abstract Coastal cliffs are dangerous because of their height and steepness, and because of rock falls or landslides. People have been killed or injured in cliff accidents. Some accidents occur when people standing or walking along the crest of a cliff fall over the edge; others when people fall when trying to climb up or down a cliff; others when people fall, or are carried down, from the top of a cliff when it subsides or collapses; and others when people are hit by falling rocks, or buried by landslides at the base of a cliff. The first two categories are due to the height and steepness of the cliff, and occur without any geomorphological change taking place; the second two are the outcome of sudden geomorphological changes. Unfortunately there are also deaths and injuries resulting from suicide or attempted suicide by leaping from a cliff, notably from high cliffs within easy reach of large urban centres, as at Beachy Head in Sussex (where there were 26 fatalities in 1990, all recorded as suicides) and The Gap on the Sydney cliffs north of Bondi Beach.

Coastal cliffs are dangerous because of their height and steepness, and because of the changes that take place, such as rock falls and landslides (Bird 1994). The number of people killed or injured by falling from cliffs is fairly small, despite the fact that several countries have public footpaths along cliffed coasts. People standing near the edge of a cliff or walking along a cliff-top footpath must be aware of the risk of falling over, a risk which increases on windy days or when visibility is reduced by heavy rain or fog. Where the cliff is receding it may be necessary to set footpaths and viewpoints back to maintain a safe distance from the cliff edge, which often requires negotiations with landowners. Cliff recession can result in the closure or diversion of coastal footpaths, and there have been cases where a coastal footpath has been diverted inland as the result of unacceptable cliff hazards or disputes with landowners. Accidents from cliff-top breakaways are unusual, but walkers should be aware of the hazard that develops when footpaths worn clear of vegetation by trampling become tension cracks near the cliff crest, along which subsidence may occur.

© The Author(s) 2016

E. Bird, *Coastal Cliffs: Morphology and Management*,
SpringerBriefs in Earth Sciences, DOI 10.1007/978-3-319-29084-3_5

Anglers sometimes fish from cliff crests (Fig. 5.1). Local councils have put up warning notices and inserted fences along particularly hazardous sections, a requirement to counter legal liability along coastal footpaths and at viewpoints (Fig. 5.2). On the French coast at Longues-sur-Mer there is a major landslip with tumbled land fronting a limestone cliff, and a notice at the cliff top says (translated): "Attention! Tourists be careful. Major risk of slumping. Do not approach the top or the bottom of the cliff—erosion has made it dangerous. The local council is not liable for accidents". In South Australia a man was killed when he fell from the cliff overlooking Maslin Beach near Adelaide, a designated nudist beach. He was apparently admiring the view.

Accidents that occur when people try to climb up or down cliffs are much commoner. In many cases climbing accidents occurred when people were attempting to climb a steep cliff when their escape along the shore had been cut off by a rising tide or a stormy sea. Others were looking for a way to or from a secluded beach, or a short cut up or down from the shore. There have been accidents to people scaling cliffs to collect eggs from bird's nests. Some people encountered problems when climbing for pleasure, particularly on coastal cliffs recognised as

Fig. 5.1 Angler fishing from a Port Campbell cliff (arrowed), Victoria, Australia. © Geostudies

Fig. 5.2 Cliff hazard warning, Mount Eliza, Victoria, Australia. © Geostudies

mountaineering challenges. Unfortunately some rock outcrops that look solid and scaleable can prove surprisingly friable when an attempt is made to climb them. Sudden rock falls and slumping occur naturally on cliffs, and may result in injuries or fatalities when people are standing or moving on the cliff at the time. In January 1990 two people who had walked out across the natural arch at London Bridge on the coast of Victoria were surprised when it collapsed into the sea behind them. They were stranded for several hours on the outlying stack before they were rescued by police helicopter (Bird 1990).

Accidents where people were killed or injured by rock falls or slumping cliffs are the result of cliff instability. As has been noted, this is greatest where the cliffs are cut in soft sandstone or clay, where the dip of the rock strata is seaward, and where there are dislocations along joints, faults or bedding-planes. The risk is lower where the cliff base is protected by a high and wide beach or a sea wall, but there is still a chance of rock falls where an indurated cliff face is suddenly breached, or where a collapse occurs below a cliff-top breakaway. A fatal accident occurred in 1990 below the Beaumaris cliffs, near Melbourne, Australia, when sandstone collapsed from a joint plane (Bird 1990). Cliff falls may also be triggered by earthquakes (Fig. 3.4), severe storms, exceptionally wet weather leading to groundwater saturation, or freezing and thawing during a cold winter. Erosion may also be caused by overloading from the weight of cliff-top buildings and by increases in seepage and runoff that usually accompany coastal urban development (Emery and Kuhn 1982).

The impact of hammering and excavation of cliff faces by people hunting for fossils or mineral specimens has resulted in slumping along cliffs near Charmouth in Dorset which are famous for their Jurassic fossils, including occasional finds of giant reptiles such as *Ichthyosaurus*. In 1977 a school group visiting Lulworth Cove in Dorset was digging for fossils at the base of a cliff that suddenly collapsed, with hundreds of tons of limestone, shale and clay falling on to the shore, killing the teacher and one of her students.

At Beaumaris on the NE coast of Port Phillip Bay, Victoria, a rich fossiliferous site occurs where Miocene marls outcrop in the base of cliffs along the axis of a

monocline. Fossil-hunters excavated a furrow in the cliff, increasing the risk of a landslide (Bird 1987). Accidents resulting from these activities have led to restrictions on the excavation of cliff faces by collectors in search of fossils or minerals.

Many rock falls occur at night, and are heard rather than seen. There are usually preliminary warnings such as falling stones, cracking as rocks move apart or snapping of plant roots, which may persuade people to retreat, clear of a landslide. In 1990 a woman and her 7-year-old son crawled to the cliff edge at Covehithe in Suffolk to look down at a stormy sea when the cliff collapsed, and they rode down on a raft of turf to the shore. Unable to climb back through soft wet mud, or to escape along the shore through large breaking waves, they were eventually rescued by a coastguard helicopter.

Warning notices and diversionary fences have been placed on or below cliffs considered to be hazardous (Fig. 5.2), partly to counter risks of legal liability and insurance problems. There is sometimes a suggestion that a hazardous cliff should be landscaped to a more gradually sloping bluff, stabilised with planted vegetation, but this results in losses of scenic and scientific values. Many such cliffs are of educational value, as well as an attraction for tourism and recreation: people want access to them, but have to be aware of the risks. In 1973 a proposal to landscape receding cliffs at Black Rock Point on the coast of Port Phillip Bay, Australia, was abandoned after protests from people who wanted the cliffs preserved as an element of scenic variety (most of the other cliffs on this coast had already been converted to grassy bluffs, Fig. 5.3) and scientists who used them for geological and geomorphological teaching and research (Bird et al. 1973). They are now listed as a Site of Scientific Interest.

Perception of risk varies between individuals and social groups. It seems likely that people who live in coastal regions are more aware of cliff hazards than those who come from inland to spend only a day or a week at the seaside. On the Glamorgan Heritage Coast in South Wales Williams and Williams (1988) found considerable variation in public perceptions of cliff hazards, with many people choosing to take risks even when they had seen warning signs. Often people

Fig. 5.3 A former cliff stabilised as a sloping grassy bluff at Quiet Corner, Victoria, Australia. Most of the natural cliff in the coastal suburb of Bayside, Melbourne, have been treated in this way: Red Bluff (Fig. 6.2) is one of the few short stretches of natural cliff on the coast of Port Phillip Bay that have survived. © Geostudies

visiting the beach sat close to the base of the cliff where there was shade from the sun. There were varying responses to different kinds of cliff hazard warning signs, those incorporating both pictorial and written information being most effective (Williams and Williams 1991); the use of international symbols avoids language problems.

Geologists face a difficult problem, for geological research and training requires access to sites on coastal cliffs where geological formations are exposed and can be studied. Botanists and zoologists also use cliffs sites for teaching and research because of their specialised flora and fauna. In Britain the Geologists' Association advises its members to consult the Coastguard Service whenever possible to learn of local hazards such as unstable cliffs or the times of high tides, or severe weather forecasts that might jeopardise excursions possible at other times. They are asked not to take risks on insecure cliffs or rock faces, and avoid dislodging rocks that might hit people below. The British government's Health and Safety at Work Act (1974, 1994) requires people visiting quarries to wear hard hats, boots and goggles, and geologists are expected to be thus attired when they visit coastal cliffs.

References

Bird ECF (1987) Geology and landforms of Beach Park, Sandringham. Sandringham Environmental Series, 2, 4th edn

Bird ECF (1990) Cliff instability on the Victorian coast. Victorian Naturalist 107:86–96

Bird ECF (1994) Cliff hazards and coastal management. J Coast Res 12:299–309

Bird ECF, Cullen PW, Rosengren NJ (1973) Conservation problems at Black Rock point. Victorian Nat 90:240–247

Emery KO, Kuhn GG (1982) Sea cliffs: their processes, profiles and classification. Bull, Geol Soc Am 93:554–644

Williams MJ, Williams AT (1988) The perception of, and adjustment to, rockfall hazards along the Glamorgan Heritage Coast, Wales. Ocean and Shoreline Manag 11:319–338

Williams AT, Williams MJ (1991) The perceived effectiveness of coastal warning signs. Coastal Zone 91

Chapter 6
Scenic and Scientific Values

Abstract Many cliffs are valued as components of coastal scenery. They provide opportunities for coastal recreation, and are of scientific interest. Problems arise where cliffs are receding, threatening coastal structures and property, and some want them stabilised, and there can be conflicts between people who want scenery preserved and access restricted and those who want to see further development. Reserves may be declared to protect scenic features or sites of scientific importance.

Cliffs are abrupt elements in coastal scenery, and are often of interest for the colour and texture of the rock formations exposed. Some serve as landmarks for sailors at sea. Artists and photographers have often chosen cliff scenes, some of which (such as the white Chalk cliffs of Dover) are well known nationally and internationally. Many cliffs are regarded as tourist attractions, such as those on the coast of Port Campbell in Victoria, Australia (Fig. 6.1). Some attract climbers and abseilers, and are used as launching places for hang-gliders, as at Stanwell Park on the coast of New South Wales. Others are important scientific sites for geologists, botanists, and particularly ornithologists (Sterry and Cleave 2012; Ellis and McLeod 2003). Coastal scenery is valued by artists and photographers, walkers and people who simply come to enjoy the view.

The need to protect coastal areas of great scenic or scientific value has led to the establishment of National Parks, Nature Reserves and various kinds of conservation sites on coasts around the world. National Parks, intended primarily to preserve coastal scenery, were originally designated in England and Wales under the National Parks and Access to the Countryside Act of 1949, and have increased subsequently. The Snowdonia, Pembrokeshire Coast, Exmoor, South Downs and North York Moors National Parks all include coastal cliffs. Areas of Outstanding Natural Beauty have also been designated, and include the Carboniferous Limestone cliffs of the Gower Peninsula in South Wales, the cliffs of South Devon between Plymouth and Torbay, and the cliffed coasts of Northumberland.

© The Author(s) 2016

E. Bird, *Coastal Cliffs: Morphology and Management*,
SpringerBriefs in Earth Sciences, DOI 10.1007/978-3-319-29084-3_6

Fig. 6.1 Cliffs and stacks in Miocene siltstone in the Port Campbell National Park, Victoria, Australia. © Geostudies

The National Trust (UK) owns numerous coastal cliff sites, particularly in Cornwall, Devon and Dorset. National Nature Reserves were set up by the Nature Conservancy from 1949 onwards, and are now managed by Natural England: there are 224 of them, established to protect habitats, species and geology, and to provide outdoor laboratories for research. Those with coastal cliffs include the Axmouth-Lyme Regis Undercliffs in Devon, Portland Bill in Dorset and Beachy Head in Sussex. Britain also has Local Nature Reserves, generally set up and managed by local authorities or County Naturalists' Trusts. Ornithological sites managed by the Royal Society for the Protection of Birds are mentioned below.

Preservation of scenic values is sometimes difficult where cliff recession is rapid, threatening coastal development, and sea walls or boulder ramparts are constructed along the cliff base. Where the rock formations exposed are coherent and resistant the cliff can be maintained, but where they are soft and friable the armoured cliff may have to be converted to a stabilised, vegetated coastal slope which is less interesting and attractive than the natural cliff. In Melbourne, Australia, there is regret that some of the cliff scenes painted by 19th century impressionists such as Tom Roberts have disappeared because of stabilisation.

Cliffs are of scientific interest because of their geomorphology, including the processes that have been discussed in this book, and their geology. They provide evidence of geological structure and stratigraphy. In Britain there are cliff exposures of rock formations in each of the geological periods, ranging from Pre-Cambrian to Pleistocene and Holocene, but not the Miocene, which is poorly represented in British geology. Certain cliff sections have been registered as 'type locations' where particular geological formations have been described and named (Fig. 6.2). Where cliffs are laterally extensive they may show relationships between rock formations that can otherwise only be deduced from inland outcrops and borings. For example the cliffs between Point Addis and Jan Juc on the coast of Victoria, Australia, show a horizon with a lateral change in rock type, the Point Addis Limestone passing eastward through calcareous clay to the Jan Juc Marl, representing an Oligocene phase when a clear sea at Point Addis passed to a muddy estuary at Jan Juc, 5 kilometres to the east (Bird 1993).

Fig. 6.2 Red Bluff, on the coast of Port Phillip Bay, Australia, is a Site of Special Scientific Interest, the type section of the Red Bluff Sand (Pliocene) over Beaumaris Sandstone (Miocene). © Geostudies

Cliffs provide access to sites with fossils representative of geological stages, and stratigraphic evidence of changing geological environments. Geological formations of Jurassic and Cretaceous age exposed in cliffs on the Dorset coast, ranging from the Lias of Lyme Regis to the Purbeck Beds near Swanage, contain deposits indicative of environmental evolution during that period, including successive marine transgressions and regressions. They include the famous dinosaur fossil sites on the 'Jurassic coast', a World Heritage site (declared by UNESCO in 2001) on the Dorset and Devon coast; a 155 km sector of coastline that also includes cliffs cut in Triassic and Cretaceous rock formations.

Reference has been made (Chap. 5) to the problems that arise when a fossil-rich site attracts intensive digging and scraping of a cliff, increasing instability and hazards. It has been necessary for scientists to remain secretive about the precise locations of cliff sites containing dinosaur fossils in the Cretaceous sandstones of the Otways and Gippsland coasts in Victoria, Australia, because important palaeontological evidence may be lost if amateur collectors excavate them.

Cliffs also expose structures indicative of tectonic history, including folds associated with mountain-building, overthrusts formed during collisions of continental plates, igneous intrusions and extrusions. Dolor Point near Coverack on the Lizard coast of south Cornwall is a cliff exposing intersecting black dykes of olivine basalt intersecting gabbro dykes in a serpentine matrix.

At Point Henry, near Geelong in Victoria, Australia a rare cliff section in Pleistocene sediments was obliterated by dumping boulders and forming a concrete ramp in order to protect a coast road.

While steep and rapidly retreating cliffs may be bare of vegetation, many cliffs have ledges and crevices where plants can become established. Most are fairly common, delivered to these habitats as wind-borne seeds or carried there by birds, but rare species may survive in these inaccessible refuges. Shakespeare referred to the 'dreadful trade' of gathering rock samphire from halfway down a Chalk cliff (now known as Shakespeare Cliff, near Dover). Slumping on cliff faces produces habitats that can be colonised by various plants, and landslides may become well vegetated. An example of this is the Axmouth-Lyme Regis landslide on the coast of Devon, where major subsidence in 1839 produced a broad irregular tumbled

undercliff backed by a Chalk upper cliff. The subsided land now carries dense forest and scrub, within which continuing minor slumping forms new habitats for colonising vegetation.

Many cliffs provide habitats for nesting and roosting seabirds and where these are particularly numerous they attract ornithologists, photographers and artists. Fortunately the collecting of eggs from nests on cliffs, a hazardous activity, has declined in Britain since it became illegal under the Protection of Birds Act 1954 (it is also illegal to possess or control any wild birds' eggs taken since that time under the Wildlife and Countryside Act 1981). A few still risk sentences of £5000 fine and/or six month's imprisonment per egg taken.

Management of cliff sites where birds are numerous seeks to minimise disturbance by activities such as rock climbing. It also provides and maintains viewpoints, including platforms from which the birds can be seen without being disturbed (Fig. 6.4). The Royal Society for the Protection of Birds (RSPB) has over a million members, and manages major cliff reserves including Bempton (Fig. 6.3) on Flamborough Head in Yorkshire (Chalk), Fowlsheugh in Kincardineshire, Scotland (Old Red Sandstone), South Stack Cliffs on Holy Island, Anglesey (Pre-Cambrian Quartzite) and St Bees Head in Cumbria (Triassic Sandstone). The preferred sites are where the cliffs are high and vertical, the rocks resistant with numerous ledges and crevices, and the shore difficult of public

Fig. 6.3 The Chalk cliff at Bempton in Yorkshire has a large colony of seabirds in a Reserve managed by the Royal Society for the Protection of Birds. The cliff-top is fenced, and the aim of management is to minimise disturbance to the birds while providing viewpoints from which they can be observed and photographed. © Geostudies

Fig. 6.4 Bird-watching platform above the cliff at Bempton. © Geostudies

access, minimising disturbance. RSPB reserve management seeks to maintain or increase the numbers of birds and variety of species, and to provide for the safety and satisfaction of birdwatchers.

Among the major bird-watching cliff sites around the world are Låtrabjarg in Iceland, Bleiksøy Island in Norway, Coburg Island in Arctic Canada and Cape Kidnappers in New Zealand.

In southern Australia colonies of the Little Penguin (Eudyptula minor) have established on cliffs as well as in dunes. In the south-west of Phillip Island, Victoria, these penguins have burrows on cliffs and bluffs of weathered basalt, and make their way each dusk up the coastal slope to burrows where they lay eggs and rear and feed their young. In the vicinity of the nightly Penguin Parade at Summerland on Phillip Island management includes the fencing of cliffs and areas where these birds have burrowed in weathered basaltic clay.

Mutton birds (short-tailed shearwater, *Puffinus tenuirostris*) have similar cliff rookeries in this region (Bird 1993).

References

Bird ECF (1993) The Coast of Victoria. Melbourne University Press

Ellis NV, McLeod, CR (2003) Sea cliffs as biological SSSIs and special areas of conservation. In: May VI, Hansom JD (eds) Coastal geomorphology of Great Britain. Joint Nature Conservation Committee: 51–53

Sterry P, Cleave A (2012) British coastal wildlife. Collins, London

Chapter 7
Stabilisation of Cliffs

Abstract Where coastal cliff recession threatens to undermine and destroy structures such as roads and buildings, or land that is valued for agriculture, or the preservation of scenic or scientific features, the usual response is to armour the cliff base against wave attack. Some cliffs have been stabilised by walling, or landscaped as bluffs. Alternatively, cliffs may be protected by beach renourishment.

The most common method of stabilising a coastal cliff is to dump boulders, concrete tetrapods, or rock ramparts to prevent waves breaking against the base of the cliff (Figs. 7.1 and 7.2). Sea walls of various kinds have been constructed to protect the cliff from wave attack.

Some cliffs are protected naturally by a wide or high beach, but where this is inadequate the existing beach may be re-shaped by bulldozing sand or shingle up against the cliff base (Fig. 7.3). Beach renourishment has been used to form a wide and high protective beach that absorbs wave energy and prevents waves attacking the cliff base (Bird and Lewis 2014). Slumping cliffs of Eocene London Clay have been landscaped and planted with vegetation, as on the north coast of the Isle of Sheppey (Hutchinson 1971) (Fig. 7.4), and attempts to stabilise landslide-prone cliffs include artificial slope grading, the insertion of drainage systems and the planting of vegetation, the most satisfactory results being where a combination of these procedures is used. In extreme cases the cliff has been stabilised by forming an artificial slope of concrete (Fig. 7.5).

Williams and Davies (1980) described the blasting of vertical cliffs of Lias limestone and shale at Lantwit Major, South Wales to produce a protective apron of rocky debris at the cliff base, but this was only a short-term solution to the problem of cliff retreat, which resumed as soon as the debris apron was washed away by wave action.

During a storm surge in Port Phillip Bay, Australia in 1935 there was severe erosion along cliffs of soft sandstone on the north-eastern coast and it was realised

© The Author(s) 2016

E. Bird, *Coastal Cliffs: Morphology and Management*,
SpringerBriefs in Earth Sciences, DOI 10.1007/978-3-319-29084-3_7

Fig. 7.1 Sea wall protecting a cliff at Black Rock, Port Phillip Bay, Australia. © Geostudies

Fig. 7.2 Concrete blocks protecting a cliff in Japan. © Geostudies

that coastal roads and buildings could be undermined by continuing cliff recession (Mackenzie 1939). The Victorian state government commissioned a survey that determined sectors that had receded by one foot or more during this storm surge, and a programme of cliff stabilisation was initiated in 1936. Masonry sea walls were built in front of vertical cliffs of soft Red Bluff (Pliocene) Sandstone, which were graded to slopes of about 25° and planted with vegetation (Figs. 7.6 and 7.7).

Where cliffs have been landscaped to sloping bluffs and planted with vegetation there may still be instability as the result of seepage. Intermittent slumping still occurs on the artificial bluffs on the north-east coast of Port Phillip Bay, notably

Fig. 7.3 Bulldozed beach at Dunwich on the Suffolk coast. © Geostudies

Fig. 7.4 A slumping clay cliff on the north coast of the Isle of Sheppey, Kent, has been largely stabilised, but there is still local erosion. © Geostudies

at Mentone, where vegetated bluffs of Tertiary sandstone became unstable when vegetation was cleared from the cliff-top area. This is because the cliff-top vegetation canopy had intercepted some rainfall, and root systems drew upon water percolating down through the soil. When the cliff-top vegetation was cleared runoff and seepage increased, causing erosion of the bluff. If groundwater becomes concentrated on a particular sector, for example where runoff from a road or car park is confined by a drain, increased slumping is likely, and saturation of fine-grained sedimentary formations can lead to collapse. Drainage schemes can reduce the risk of slumping, as can the maintenance of a dense vegetation cover on a coastal slope, which intercepts rainfall and impedes runoff, and absorbs seepage.

A major example of cliff stabilisation is seen at Lyme Regis in Dorset, where Church Cliffs had been the scene of recurrent slumping in Lower Lias clays and limestone layers. Insertion of drainage pipes had been used to reduce seepage by intercepting groundwater and conveying it to outfalls in order to halt slumping,

Fig. 7.5 Recession of a sector of Chalk cliff on the Isle of Thanet, Kent, halted by concrete walling. © Geostudies

but in 2014 a more elaborate concrete sea wall was built along the base of the cliffs, and the coastal slope graded, with drainage pipes, pins and netting inserted to keep it stable, and planted with grassy vegetation (Fig. 7.8). On some cliffed coasts attempts have been made to stabilise a cliff by inserting cables and anchors, fitted into backing solid rock, an adaptation of techniques used to retain walls and buildings at risk of collapse. This has generally been more successful on roadside cuttings than on open coasts.

Stabilisation of coastal slopes cut in soft sediments may be achieved by establishing a vegetation cover. This is often achieved by placing a blanket of biodegradable material over the slope and planting grasses or shrubs that develop into protective vegetation (Fig. 7.9). The grassy slope at Quiet Corner, Victoria (Fig. 7.7) was stabilised after a vertical cliff in soft sandstone had been graded to a slope of 30° then planted with *Stipa* grass tussocks, held in place by scattered rocks.

Where cliff recession threatens to undermine and destroy roads and houses, as on the Holderness coast in eastern England, there is a demand for stabilisation, often at a cost that exceeds the value of the threatened property. The alternative is 'managed retreat', in effect abandoning the threatened structures and replacing them inland. A corollary of this is the prohibition of structural development

Fig. 7.6 Artificial concrete slope fronting a house on the coast of Ecuador. Erosion continues on adjacent sectors of sloping clay cliff. © Geostudies

Fig. 7.7 Former vertical cliffs were graded and stabilised behind a sea wall at Black Rock, Port Phillip Bay, Australia, in 1939, but the beach disappeared as the result of wave reflection from the sea wall. © Geostudies

Fig. 7.8 The stabilised coast at Church Cliffs, Lyme Regis, Dorset, in 2015. © Geostudies

Fig. 7.9 Stabilisation of a coastal slope in sandy clay at Santa Catterina di Pittinuri on the west coast of Sardinia. © Geostudies

on land expected to be lost as the result of cliff recession. Evidence of rates of cliff recession can be used to draw set-back lines to indicate where the coastline is likely to be in 50 or 100 years time, taking account of sea level rise predictions. The extent of damage to coastal structures in stormy years in Britain has prompted a demand that the government spends massively on coast protection works. After a stormy winter in 2015 the British government announced that it would spend £3.2 billion on flood management and coastal protection for 15,000 houses. However in lengthy sectors of rapidly receding coast such as Holderness coast protection may prove economically unsustainable.

References

Bird ECF, Lewis N (2014) Beach renourishment. Springer, Dordrecht
Hutchinson JN (1971) Surveys of coastal landslides: Kent. Building Research Station, Watford
Mackenzie AD (1939) Coastal erosion in Victoria. Transactions of the Institute of Engineers, Australia 20:229–236
Williams AT, Davies P (1980) Man as a geological agent: the sea cliffs of Llantwit Major, Wales. Z Geomorphol Supplementband 34:129–141

Chapter 8
Summary

Abstract Coastal cliff management should be based on an understanding of geomorphology: how a cliff or bluff has formed, how it is changing naturally and as the result of human activities, and how management procedures are likely to modify it. This requires knowledge of the processes at work, their effects, and changes likely to occur if they are modified.

Cliffs are the outcome of erosion, past and present, and surveys are necessary to determine their evolution, rates of change, and future development in relation to coastal processes. They can be stabilised, modified, or managed in ways that conserve their scenic and scientific values and recreational use. It is not desirable, even if it were technically possible, to stabilise all cliffs and thereby halt coastal evolution. One should be wary of the attitude of some engineers who see cliffs as features to be beautified with concrete and boulders. Natural cliff recession maintains an interesting and varied environment, providing habitats for wildlife and yielding sediment to supply nearby beaches. It is worth recalling that coastal cliffs are, and have long been, natural features in the landscape, and that cliff recession has occurred throughout geological time,. It should not be assumed that cliff erosion is necessarily a problem requiring a cure. Already many cliffs have been walled up or 'landscaped', sometimes unnecessarily, and often followed by the erosion of adjacent or nearby beaches.

It is necessary to avoid procedures that have been shown to lead to cliff instability. These include coastal quarrying, the use of explosives on cliffs, and the depletion of beaches that previously protected cliffs from wave attack. Instability can be caused by the making of roads or increasing the load of buildings on or upon cliffs. Corniche roads on cliffs or bluffs can cause slope instability during and after their construction; they are expensive to maintain and generate hazardous traffic.

Cliffs are damaged by people scrambling up or down them, and by people hunting for fossils or minerals. Runoff and seepage result in cliff erosion, and can be controlled by inserting cliff-top drains or piping to control groundwater flow.

© The Author(s) 2016

E. Bird, *Coastal Cliffs: Morphology and Management*,
SpringerBriefs in Earth Sciences, DOI 10.1007/978-3-319-29084-3_8

Vegetation is often an important factor in maintaining the stability of coastal slopes, and can be introduced or improved to counter erosion. Where vegetation is stabilising a coastal slope it should be carefully conserved. Depletion of vegetation along cliff-top footpaths may result in breakaways and cliff recession.

Cliff stabilisation by building sea walls or similar structures may be necessary to prevent damage to roads, buildings and valuable land on cliff tops, but the scenic and scientific values of natural cliffs should be conserved where possible.

Glossary

Abrasion the wearing away of a rock surface by friction as the result of the impact of winds, waves or currents, particularly when these agents are armed with rock fragments (sand or gravel); sometimes called corrasion

Alpine earth movement the folding and faulting of rock formations as the result of stresses associated with the uplift of the Alps in Europe in mid-Tertiary times; extending to southern England including the Weald the London Basin, the Isle of Wight and Dorset

Alum aluminium and potassium sulphate, a mineral which has household and industrial uses

Anticline an upfold or arch in rock strata

Basalt a dark, fine-grained basic igneous rock produced by volcanic activity

Beach renourishment the restoration of a depleted beach by depositing sand or gravel

Bedding plane the contact between layers of rock strata: see Fig. 2.17

Bioerosion erosion of a rock surface by physical and chemical processes associated with the activities of plants or animals: see Figs. 3.23, 3.24 and 3.25

Blowhole a hole or fissure in the roof of a cave through which fountains of water and spray are forced by intermittent air pressure trapped by incoming waves

Bluff a steep coastal slope stabilised beneath a soil and vegetation cover. In North America the term is often used as a synonym for cliff

Boulder Clay a mixture of clay and stones deposited by a glacier

Breakaway a fracture on a cliff behind an area where rock has subsided

© The Author(s) 2016
E. Bird, *Coastal Cliffs: Morphology and Management*,
SpringerBriefs in Earth Sciences, DOI 10.1007/978-3-319-29084-3

Brickearth a fine-grained silty sediment deposited by wind action; also known as loess

Buttress a narrow vertical rock projection from a cliff: see Fig. 2.30

Calcareous containing calcium compounds, notably calcium carbonate

Calcareous siltstone a fine-grained calcareous sediment coherent enough to stand as a cliff: see Fig. 3.2

Calcilutite a calcareous siltstone, q.v

Cambrian the earliest period (510–560 million years ago) in the Palaeozoic era

Caprock a layer of resistant rock on top of a cliff

Carboniferous a period 290–360 million years ago in the Palaeozoic era

Case hardening formation of a harder superficial crust on a rock outcrop by precipitation of cementing material such as carbonates or iron oxides

Castellated a landform like a castle wall, with battlements

Cave a natural enclave in a cliff

Cavitation the formation of a cave or cavity in a rock surface

Chalk a soft or moderately hard white or grey limestone composed largely or entirely of calcium carbonate; the upper division of the Cretaceous period

Cleavage parallel planes of division in a rock that splits into thin sheets (such as slates)

Cliff a steep slope exposing rock formations

Condominium a large building containing many apartments

Corniche a road along a steep coast: see Fig. 3.31

Cretaceous the last period of the Mesozoic era, 65–144 million years ago

Cyclone a violent storm around a centre of low atmospheric pressure; also known as a hurricane or typhoon

Debris fan accumulation of disintegrated rock material at the base of a slope

Devonian a period between 360 and 405 million years ago, during the Palaeozoic era

Dune calcarenite a generally consolidated aeolian (wind-deposited) sandstone lithified by cementation or partial cementation of dune sand by secondary internal precipitation of carbonates from groundwater: see Fig. 3.55

Dyke a long and narrow intrusion of volcanic rock

Earth cliff a cliff composed of soft rock

Eocene an epoch 36–53 million years ago, during the Tertiary period

Escarpment cliff a coastal cliff cut across rock formations that are horizontal or dipping landward

Fault cliff a cliff produced by land uplift along a fault line

Fault line the outcrop of a fracture in the land produced by displacement of rock formations (faulting)

Fault-line cliff a cliff cut by differential erosion along the line of a fault that had juxtaposed rock formations of contrasted resistance, so that the cliff exposes the fault plane

Ferruginous containing iron compounds

Ferromagnesian containing iron and magnesium compounds

Fetch the length of open water across which waves are generated by shoreward wind action

Frost shattering the disintegration of rock by recurrent freezing and thawing

Gabbro a basic crystalline igneous rock

Gault a thick heavy clay formation between the Upper and Lower Greensand in the Cretaceous rocks

Geomorphology the shaping of landforms

Glacial drift deposits left by a melting glacier

Glauconite a ferromagnesian clay mineral chiefly found in marine sediments

Gneiss a coarse crystalline metamorphic rock

Goethite a ferric iron oxide common in ferruginous sandstones

Granite a crystalline igneous rock containing quartz, feldspars and mica

Greenstone a dark green metamorphic rock, originally dolerite, gabbro or basalt, hardened by heat and pressure

Gulleying formation of deep, narrow incised channels by runoff on cliffs or steep slopes: see Fig. 3.7

Haematite a form of ferric oxide common in ferruginous sandstones

Hawkesbury Sandstone a Triassic sandstone formation extensive in Central New South Wales and prominent on the Sydney coast: see Fig. 2.33

Head deposit a poorly sorted angular rubble in an earthy matrix formed by periglacial (freeze-and-thaw) weathering and extensive on slopes in periglaciated areas such as SW England

Hogsback cliff rounded, convex coastal slopes in previously periglaciated areas, notably on the North Devon coast: see Fig. 2.41

Holocene the last 10,000 years of geological time

Humate a consolidated deposit formed within dunes as a subsoil horizon enriched by downwashed iron oxides and organic matter leached from overlying sand, or a compacted peaty swamp deposit: see Fig. 4.1

Igneous rock formed where molten rock (magma) from the Earth's interior has been extruded on to the surface (volcanic rock) or intruded into the crust to cool as crystalline rock (e.g. granite)

Induration hardening of a rock formation by precipitation of cementing materials (carbonates, iron oxides, silica) or by heat and pressure, as in metamorphic rocks

Isostatic depression of the Earth's crust as the result of loading of the surface by sediment accumulation, volcanic deposition or ice accumulation, and uplift as the result of unloading, as when an ice sheet melts (postglacial isostatic recovery)

Joints vertical, horizontal or inclined planes of division in a rock formation formed after consolidation (shrinkage of cooling igneous rocks or drying sedimentary rocks)

Jurassic a period between 144 and 213 million years ago in the Mesozoic era

Jurassic Coast the name given to the south coast of England between Orcombe Point in Devon and Old Harry Rocks in Dorset, which has cliffs cut in Jurassic formations and also Triassic and Cretaceous formations; it is a World Heritage site

Kaolinite a soft white clay mineral formed by the decomposition of feldspars

Karst landforms shaped by solution processes, notably on limestones karstic features include surface depressions sinkholes, caves and irregular outcrops (lapiés)

Landslide mass movement of rock or rubble down a slope; also known as a landslip: see Fig. 3.15

Lapiés etched, pitted and spiky surfaces formed by solution on limestones: see Fig. 3.22

Laterisation weathering of rocks under humid tropical conditions to form clay minerals containing oxides of aluminium, iron and manganese

Laterite weathered rock produced by laterisation q.v

Lias the lowest division of the Jurassic rock formations

Lithology the characteristics of a rock, including its mineral composition, structure and grain size

Litorina Sea the name given to a marine transgression that occurred in the Baltic region around 8000–6000 years ago, followed by a fall in sea level that produced the marine foreland (see Fig. 2.37)

London Clay a clay formation of Eocene age, extensive in the London Basin

Marine transgression a rising sea, relative to the land

Massive rock in which planes of division (joints or bedding-planes) are widely spaced

Megacliff a cliff more than 500 m high: see Fig. 1.2

Metamorphic rock rock that was originally igneous or sedimentary, which has been altered by heat, pressure or permeating liquids or gases, producing new structures and minerals

Microcliff a cliff generally less than a metre high

Miocene an epoch 5 to 23 million years ago in the Tertiary period

Monocline a sharp flexure between horizontal rock strata at different levels

Montmorillonite a clay mineral that swells in volume because of water absorption, and shrinks as it dries out

Moraine a ridge of sediment (silt, sand, gravel, boulders) deposited by a melting glacier

Mudrock a massive rock composed of indurated fine-grained sediment (silt and clay)

Mudflow where highly lubricated fine-grained sediment moves downslope as a slurry

Mudslide where a mass of coherent silty or sandy clay moves irregularly down a slope

Natural arch a tunnel excavated through a headland, island or stack: see Fig. 2.27

Notch and visor profile a hollow excavated along the base of a cliff with an overhanging rock projection, see Fig. 2.6

Ocean swell regular waves transmitted across an ocean from a stormy area

Old Red Sandstone a sandstone of Devonian age

Olivine a green mineral common in basic volcanic lava, notably basalt

Palaeontological pertaining to fossils

Palaeozoic the geological era including Cambrian to Permian, 560–248 million years ago

Peat partly decomposed swamp vegetation

Periglacial process or environment bordering the limits of glaciation, typified by recurrent freezing and thawing, including snowfall and snowmelt

Permafrost permanently frozen ground

Phyllite a metamorphosed fine-grained rock formed by the intense compression of slate with crumpled cleavage and silky texture

Pindan a type of scrub woodland vegetation in NW Australia, and also the red silty clay formation on which it grows: see Fig. 2.19

Pleistocene the geological epoch 10,000–2 million years ago, preceding the Holocene in the Quaternary period

Plunging cliff a cliff or steep coast that descends into deep water without any intervening shore platform, rocky shore or beach: see Fig. 1.2

Portland Limestone a limestone near the top of the Jurassic rock sequence on the Dorset coast: see Fig. 2.48

Postglacial the period following the melting of Pleistocene glaciers and ice sheets

Pre-Cambrian the geological era preceding the Palaeozoic era, more than 560 million years ago

Purbeck Beds a rock formation at the top of the Jurassic rock sequence on the Dorset coast

Quartzite a rock formation consisting largely or entirely of quartz (silica), occurring either as an igneous intrusion (dyke) or indurated sandstone

Quaternary the geological period that began about 2 million years ago, and comprises the Pleistocene and Holocene epochs

Raised beach a beach that has been uplifted by tectonic movements to stand above the level at which it originally formed

Rill a small narrow channel formed by runoff

Rock fall the collapse of a rock or rocks from the face of a cliff; also termed a cliff fall: see Fig. 3.35

Rotational slide a landslide down a curved slip plane, forming a back-tilted rock mass

Runoff the flow of water down a slope

Salt plucking the disintegration of a rock surface caused by the growth of salt crystallising from sea spray

Sedimentary rock rock formed by the deposition of sediment in water or on land

Seepage the flow of water out of a rock formation

Shear strength resistance to shearing (the fracturing of rock)

Shear stress the cause of shearing (the fracturing of rock)

Shore platform a flat or gently sloping rock surface extending between high and low tide levels: see Fig. 2.1

Siliceous containing silica

Silicified indurated by the deposition (precipitation) of silica

Slope-over-wall coast a coastal profile with an upper slope descending to a steeper, often vertical cliff: see Fig. 2.49

Slumping collapse of a slope

Slurry semi-liquid mud (silt or clay)

Smectite a clay that swells in volume as it absorbs water and shrinks as it dries out

Soil creep the slow downslope movement of soil or surface sediment

Solifluction the slow downslope movement of sediment or weathered debris lubricated by waterlogging, notably when a frozen land surface thaws

Solution the dissolving of soluble rock material, notably carbonates, in water

Stack an isolated steep-sided rock pillar or column rising from the shore, a shore platform, or the sea floor near a cliffed coast: see Fig. 6.1

Storm waves produced by strong wind action during a storm

Strata layers of rock in a stratified formation

Stratigraphy study of the geological sequence and position of rock formations

Subaerial exposed to the atmosphere

Syncline a downfold in rock strata

Talus apron a slope-foot accumulation of rocky debris: see Fig. 3.4

Tectonic movements of the Earth's crust, including earthquakes; folding and faulting of rock formations

Tectonic plates migrating areas of Earth's crust on which the continents stand

Terrace a flat or gently sloping strip of land bordered above and below by steeper slopes. See Figs. 2.7 and 2.33

Tertiary the geological period between 2 and 65 million years ago, consisting of the Palaeocene, Eocene, Oligocene, Miocene and Pliocene epochs

Toppling separation and collapse of columns of rock from a cliff face: see Fig. 3.2

Translational slide where a rock mass slips down a seaward-sloping dipping plane

Triassic the period, 213–248 million years ago, at the beginning of the Mesozoic era

Tsunami a sea wave generated by a major disturbance within an ocean basin (earthquake, submarine volcanic eruption or landslide), which increases in height on entering shallow water, and can exceed 30 metres when it breaks on a coastline

Undercliff a lowland fringe below a cliff or steep coast produced by subsidence or erosion: see Fig. 2.20

Upper Greensand a rock formation immediately underlying the Chalk in the English Cretaceous

Volcanic ash fine-grained sediment ejected from a volcano, known also as tuff

Wealden alternating sandstone and clay formations in the Lower Cretaceous in England

Weathering the disintegration and decomposition of a rock surface exposed to the atmosphere as the result of physical, chemical and biological processes

MIX
Papier aus verantwortungsvollen Quellen
Paper from responsible sources
FSC® C105338

If you have any concerns about our products,
you can contact us on
ProductSafety@springernature.com

In case Publisher is established outside the EU,
the EU authorized representative is:
Springer Nature Customer Service Center GmbH
Europaplatz 3, 69115 Heidelberg, Germany

Printed by Libri Plureos GmbH
in Hamburg, Germany